AF215273

SHIPBUILDING TRENDS AND THE RISE OF THE INDO-PACIFIC

SHIPBUILDING TRENDS AND THE RISE OF THE INDO-PACIFIC

Commodore Sanjay Kumar Jha

Copyright © National Maritime Foundation, New Delhi, 2024

All rights reserved. No part of this publication may be reproduced,
stored in a retrieval system, or transmitted in any form or by any
means, electronic, mechanical, photocopying, recording or otherwise,
without the prior written permission of the Publisher.

First published in 2024 by
PENTAGON PRESS LLP
206, Peacock Lane, Shahpur Jat
New Delhi-110049, India
Contact: 011-26491568 • 011-26490600

Typeset in Adobe Garamond, 11.5 Point
Printed by Aegean Offset Printers, Greater Noida, U.P.

ISBN 978-93-90095-95-7

Disclaimer: The views expressed in this book are those of the author and
do not necessarily reflect those of the National Maritime Foundation,
or the Government of India.

www.pentagonpress.in

CONTENTS

PREFACE

After the Second World War, countries around the world began working on reconstruction and economic recovery. Many of these countries had either recently gained independence from colonial rule or had emerged at the fall of once-mighty empires. Since the post-war world order was essentially bipolar, with the United States of America (US) representing one pole and the Union of Soviet Socialist Republics (USSR or the Soviet Union) the other, most countries aligned themselves with one or the other of these great powers to facilitate their own growth. The dynamics of the global policies of the US and the USSR, involving containment on the one hand and balance-of-power on the other, defined these alignments. In furtherance of its policy of containment of the Soviet Union, the US formulated a two-pronged approach, with one prong being economic and the other military. The economic prong was manifested in the Marshall Plan – a blueprint to help restore and revive the economies of countries of Western Europe through the provision of financial aid, the transfer of advanced industrial technology, and the grant of access to markets in the US. Emblematic of the military prong was the creation of a military alliance structure – the North Atlantic Treaty Organization (NATO). The USSR responded by creating its own set of military alliances with countries that it had enveloped within its fold – the Warsaw Pact.

Although the Second World War had ravaged Europe, and engendered poverty, severely degraded agricultural growth, and dysfunctional or absent industrial infrastructure, there were nevertheless, two saving graces that had endured. First, mass production in heavy engineering had matured and was well-established. Europe's heavy engineering industries were amenable to being relocated to places that had a low-cost but skilled workforce that could be employed to increase profits. Second, the primacy of sea power (naval power) – which had been established after the Great Voyages of the sixteenth century, when sea routes, protected by navies, became the principal component of the global logistic supply chains – was intact. The US, of course, had the enormous advantage of its advanced industrial infrastructure, which had been upscaled by the demands of the Second World War. It had the highest economic reserves; and it was the leading maritime and naval power, with the most advanced shipbuilding industry.

One of the first countries outside of the Warsaw Pact to record high growth was Japan, which had been vanquished in the Second World War and placed under US control. Under the Supreme Commanders of Allied Powers (SCAP), the country implemented extensive agricultural reforms, and utilised financial aid and advanced technology from the US and its allies to transform itself into an export-oriented industrial society. Other countries in maritime Southeast Asia followed suit in the later decades of the last century, using the blueprint of the 'Japanese miracle' incorporating the transfer of technology from the West, setting up export-oriented industries, and accessing the US market as also those of the latter's allies, for profits and growth. Amongst the industries that were relocated to maritime Asia – with Japan at the helm – was the shipbuilding industry. The US had once been the world leader in shipbuilding but in the aftermath of the Second World War, it found itself with excess capacity and falling demand. On the other hand, the pre-war shipbuilding infrastructure of Japan was largely intact, and shipbuilding offered an opportunity for large-scale employment and transfer of technology since this is an inherently technology- and manpower-intensive industry. Shipbuilding, therefore, continued to be a common thread, facilitating and fuelling the economic growth and industrialisation of other countries, such as South Korea, China, and India.

The growth of countries in maritime Asia helped the region emerge as the leading manufacturing hub of the world. Consequently, countries of the region have not only become major manufacturers in the technologies of the Third Industrial Revolution (Industry 3.0) but are also contenders in the race to be in the vanguard of innovation and the development of technologies of the Fourth Industrial Revolution (Industry 4.0). As a region, Asia collectively produces more than ninety per cent of ships in the world and has now emerged as a significant contributor to the world's GDP. In the process of setting up some of the most state-of-the-art industrial infrastructure for the export of advanced shipbuilding and associated products, these countries also began to utilise them internally to produce ultra-modern warships and submarines. This was achieved in collaboration with the leading Western producers of weapon systems and warships, who helped these countries emerge as significant players in the global military-industrial complex.

The high growth recorded by maritime Asia after World War II has brought the Indo-Pacific construct to prominence as a region of intersecting interests, including those of major maritime trading and strategic powers, which significantly shape global developments. The landscape is strategically bound by India in the west (although in the future it will inevitably stretch farther westwards to incorporate at least the eastern coast of the African continent), the US in the east, Japan and South Korea in the north, and Australia in the south. This region encompasses more than half the earth's surface, supports more than half the world's population, and contributes as much to the world's GDP and is now known as the Indo-Pacific, stretching from the eastern shores of Africa to the western shore of the Americas. As countries of the Indo-Pacific record consistently high economic growth, a huge middle class is emerging – especially in the Indian Ocean segment of the Indo-Pacific (comprising the Indian Ocean and its island-States) – that has a markedly youthful profile and significant purchasing power. This facet adds to the power of this region to influence markets, since consumer data is the most potent propellant of the growth of technologies within Industry 4.0, such as artificial intelligence (AI), machine learning (ML), the Internet of Things (IoT), additive manufacturing, robotics, etc. A study of shipbuilding trends of a representative sample of countries of the region would help one to understand how the interplay of history, technology, economics, and strategy has contributed towards the growth of the Indo-Pacific – which is arguably the single most defining feature of the post-World War II period.

ACKNOWLEDGEMENTS

At the outset, I wish to acknowledge the opportunity provided to me by the National Maritime Foundation (NMF) to publish this book – *"Shipbuilding Trends and the Rise of the Indo-Pacific"*. The present work is based upon my PhD thesis, approved by the Naval War College and University of Mumbai in July 2022. Accordingly, I am grateful to the NWC and the MU for providing me with the opportunity to pursue my research into the fascinating and contemporary subject of *"Shipbuilding Trends in the Indo-Pacific: The Interplay of History, Technology, Economics, and Strategy"*.

I am indebted to my PhD Research Guide, Dr Odakkal Johnson; interim guide Dr Sanjay J Singh; and Dr Vijay Sakhuja, my guide from November 2016 to July 2020. Their in-depth experience and understanding of defence and strategic studies, along with their active supervision, insightful guidance, patience, and ready willingness to assist, have been pivotal to the undertaking and completion of my PhD.

In addition to my PhD thesis, this book draws heavily upon my 35 years of experience as a naval officer and an electrical engineer, building on earlier research undertaken by me for obtaining an MPhil (Arts) from the University of Mumbai, whilst undergoing the 19th Naval Higher Command Course (NHCC) at the Naval War College in 2006-2007. I would like to accordingly express my gratitude, and acknowledge the support and opportunity provided to me by the Naval War College. I extend my sincere thanks to Ms Krithi Ganesh – Research Associate, NMF – who edited the script, designed the book's cover, and gave the book its final shape.

I especially acknowledge and note my deepest gratitude to my wife, friend, and soulmate, Mrs Lipika Jha, for her constant support and the conducive environment that she always provided, enabling me to make many long sojourns into the academic world and cyberspace, while she managed the real world around us in an able, happy, and efficient manner. I also thank my children, Kush and Neil, for their steady encouragement and motivation.

Place: New Delhi   **(Sanjay Kumar Jha)**
Date: Aug 2023

LIST OF ABBREVIATIONS

1st IR	-	First Industrial Revolution (Industry 1.0)
2nd IR	-	Second Industrial Revolution (Industry 2.0)
3D Printing	-	3-Dimensional Printing
3rd IR	-	Third Industrial Revolution (Industry 3.0)
4th IR	-	Fourth Industrial Revolution (Industry 4.0)
AI	-	Artificial Intelligence
BCE	-	Before the Common Era
CE	-	Common Era
CPS	-	Cyber-Physical Systems
FMS	-	Foreign Military Sales
GRSE	-	Garden Reach Shipbuilders and Engineers
GSL	-	Goa Shipyard Limited
HSL	-	Hindustan Shipyard Limited
IoT	-	Internet of Things
IVI	-	Industrial Value-chain Initiative
K Waves	-	Kondratiev Waves
LNG	-	Liquefied Natural Gas
MDL	-	Mazagon Docks Limited
MIC	-	Military Industrial Complex
ML	-	Machine Learning
NATO	-	North Atlantic Treaty Organization
OPEC	-	Organization of Petroleum Exporting Countries
SCAP	-	Supreme Commanders of Allied Powers
V/AR	-	Virtual/Augmented Reality

LIST OF FIGURES AND TABLES

FIGURES

TABLES

INTRODUCTION

This book is the result of a study on the interplay of history, economics, technology, and strategy that shaped shipbuilding trends in maritime Asia in the post-World War II period. It seeks to provide greater insight into the rise of export-oriented industrial economies in this region. The book applies a multi-dimensional model of analysis of this phenomenon in order to develop a deeper understanding of the emergence of the 'Indo-Pacific' construct and attempts to understand why some countries succeeded whilst others were not as successful despite following the same blueprint for progress. The book also examines the impact of Industry 4.0[1] technologies and green technologies on shipbuilding trends and their significance in the Indo-Pacific region. It is believed that such an understanding can form the basis of effective policy formulations and facilitate growth and prosperity in many countries.

Outline of Chapters

Chapter 1: Prelude: The Rise of Maritime Asia in the Aftermath of the Second World War. This chapter introduces the 'Indo-Pacific' construct, touches upon the rise of export-oriented maritime economies in Asia in the post-World War II period, and brings out the correlation between the global trends in shipbuilding and its linkage with the rise of the 'Indo-Pacific'.

Chapter 2: Shipbuilding – A *Tour de Horizon*. This chapter examines the evolution of shipbuilding and the effect of global maritime trade and sea power. It outlines the existence of distinct patterns of global maritime trade in medieval times in the Indian Ocean region as well as the Mediterranean. It briefly narrates the manner in which the period of peaceful sailing in what we now refer to as the Indo-Pacific changed in the early sixteenth century as merchants from European nations began to venture into the Indian Ocean and beyond, ushering a new era of trade-driven competition wherein dominant maritime powers began to assert commercial and political superiority. It elucidates the interplay of history and strategy that facilitated the rise of sea power, economics, and strategy, thus leading to the modern world system, as well as the role of economics and technology in the world economic cycles, that is, the 'Kondratiev Waves'. Within this chapter, development in shipbuilding has been consolidated under four distinct epochs. The first is until the fall of Constantinople; the second is from 1500 to 1750 AD – the era of the great voyages of discovery, and the financial and scientific revolution in Europe; the third period extends from 1760 to 1850, which was the age of the industrial revolution; and finally, the fourth period is from the 1850s to the 1940s, and encompasses the Great War (popularly but inaccurately called the 'First World War') and the Second World War.

Chapter 3: Context of Technology Power-Shifts in Shipbuilding. This chapter examines the theoretical context of power-shifts over the last three centuries, with a specific focus on the resurgence of maritime Asia and the primacy of technology in facilitating the shifts of power in the latter half of the twentieth century. As the multipolar world of the 1870s transformed into the bipolar world of the 1940s – with the US pursuing containment as its strategy of choice, power transitions began to be greatly influenced by the balance of power. Different approaches adopted by the Soviet Union and the US are analysed (albeit in a truncated manner), including Mackinder's 'Heartland Theory', Spykman's 'Rimland Theory', and Mahan's 'Theory of Influence of Sea Power'. The genesis of the Marshall Plan and NATO are examined, as also the developments that led to the rise of Japan, South Korea, and China – amongst other countries in the Indo-Pacific.

Chapter 4: The Technology Conundrum: Transitions & Transformations. Chapter 4 examines the rise of countries of the Indo-Pacific as manufacturing hubs, and the consequential expansion of the middle-class, which has now become the largest consumer market. Economic and industrial progress has also helped the Asian countries increase their military power. The chapter presents a classification of the Industrial Revolution into distinct periods – Industry 1.0, Industry 2.0, Industry 3.0, and Industry 4.0 – during each of which, specific technologies substantially changed societies. Amongst the technologies and applications that are critically analysed are Artificial

Intelligence (AI), Machine Learning (ML), Virtual/Augmented Reality (V/AR), 3D Printing, Blockchain, Autonomous Vehicles, advanced materials, Digital Twinning, the Internet of Things (IoT), and Cyber-Physical Systems (CPS). Several of these technologies are reshaping the maritime world.

Chapters 5 through 8 are case studies that analyse the growth of Japan, South Korea, China, and India, with particular emphasis on the shipbuilding trends in these countries as drivers of the Indo-Pacific construct. Each case study is presented through four time periods: before the mid-19th Century; mid-19th Century to the end of World War II; and post-World War II. In the period after the Second World War, advancements in the warship-building industry and recent developments in shipbuilding have been examined for each country.

Chapter 9: Conclusion. The final chapter summarises the key findings of the evaluation undertaken in terms of the role of history, technology, economics, and strategy in defining the trends of shipbuilding in the countries in maritime Asia in the post-World War II period, and the manner in which this has contributed to the rise of the Indo-Pacific.

1

Prelude: The Rise of Maritime Asia in the Aftermath of the Second World War

The post-Second World War period saw the reconstruction and restoration of countries around the world. Major economies emerged in the Asia-Pacific region and began to influence geopolitics to an extent unseen in recent history. The full impact of this shift in the concentration of geopolitical developments in the Asia-Pacific region is only now beginning to unfold; its magnitude is so pronounced that it could potentially change the course of history – which is to say, the impact will be profoundly long-term. Another interesting facet is that this development is unfolding against the backdrop of a number of major technological transformations that range from heavy engineering and the advent of the 'Information Age', to the emerging technologies of Industry 4.0. There are obvious linkages between the factors that contribute to this trend of economic boom, the industrial development of countries of the Indo-Pacific, as also as the accompanying geopolitics. The growth story of any major industry in a country usually correlates with the overall national growth story and this, by extension, would apply to the region as well. Therefore, an examination of the global trends of any major industry will highlight the underlying facets of this development and bring out the linkages between the two. Accordingly, this author has attempted an examination of global shipbuilding trends so as to better understand the rise of the Indo-Pacific after World War II.

The growth of shipbuilding worldwide has been subjected to scholarly enquiry from a variety of perspectives, ranging from the historical and technological to the economic and grand-strategic, and yet, these have seldom been examined in an integrated manner. The evolution of ships and societies in the Indo-Pacific region merits a unified multi-dimensional model of analysis that holistically explains shipbuilding trends as an interplay of the parameters of history, economics, technology, and strategy. This book is an attempt towards the implementation of such a unified multi-dimensional model to understand how the evolution of shipbuilding has been continually shaped by a slew of factors, including those of economics; trade and commerce; technological transformations; as well as the primacy of sea power in military strategy. The book aims to use this approach while examining shipbuilding in select maritime Asian countries in the post-World War II period, thereby exposing the manner in which shipbuilding facilitated the growth of industry-based export-oriented economies that eventually led to the rise of the 'Indo-Pacific' construct.

This approach is valid for at least four reasons. First, a ship is an amalgamation of a variety of technologies – metallurgy, design, architecture, construction, machinery, electronics, and data processing systems, amongst others.[2] Second, it is a human-resource-intensive industry that employs many people – directly or indirectly – as part of a large industrial complex, making it an industry of choice for transforming an industrial society. Third, it is not only the earliest vehicle and facilitator of globalisation but continues to be the principal component of the modern world trading system.[3] An overwhelming volume of goods is transported by sea; even in the current digital age, any product ordered with the click of the mouse has a ninety per cent chance of having travelled partly or wholly on a ship. Fourth, warships and submarines produced by this industry form the backbone of a nation's sea power.

Understanding the Rise of the Indo-Pacific

The immediate aftermath of the Second World War witnessed the advent of a new world order – one that was essentially bipolar. On the one hand, it was abundantly clear that the age of empires and colonies was at an end. On the other, newly independent nation-states such as India, China, Indonesia, Philippines, Iran, Taiwan, the Baltic States, Finland, Hungary, and Romania[4] were open to influences from the established power centres, i.e., the US and the Soviet Union. These countries had to choose the type of economy that would meet their requirements, the form of governance that would be acceptable to their people, and the world power they would lean towards. They made their choices based on their respective comprehensive national interest and the compulsions dictated by the prevailing state of the economy. Since many of these newly independent States were in the continent of Asia, the ensuing period witnessed a profound influence on growth patterns in this region, which was itself a result of the recalibration of great power equations and the dynamics of the balance of power.

The post-World War II world had already been industrialised for two centuries and thus had two prominent characteristics. First, Western countries had a well-established industrial ecosystem for mass production in heavy engineering. Industrial complexes in these countries were at a 'mature' stage of their life cycles, which rendered them amenable to relocation to places where low-cost skilled workforce could be employed to maximise profits. Second, the primacy of sea power – as had been established centuries earlier – was still an essential requirement for the acquisition of great power status. Amongst the leading powers, the US had some exclusive advantages, including advanced industrial infrastructure that remained largely unaffected by World War II, the highest economic reserves, and a formidable sea power with global outreach. All these contributed significantly to the US having the world's most advanced shipbuilding industry at that point in time.

As countries in Asia began to industrialise, they embarked upon economic progress that was essentially export-oriented. Using this approach, they recorded significant economic growth in a short span of time. One of the earliest among such countries was Japan, whose economic rise happened in stages – starting with core industries like coal, steel, and shipbuilding; followed by consumer products and automobiles; and finally, knowledge-based products like computers and electronics. In the subsequent years, many countries attempted to adopt the 'Japanese miracle'. Between 1965 and 1995, South Korea, Taiwan, Hong Kong, and Singapore – often referred to as the Four Asian Tigers – saw their per capita incomes increase six-fold, while Indonesia, Malaysia, and Thailand managed to triple their income levels. China, and later India, began to record high growth by the 1980s and 1990s, respectively.

In the second half of the twentieth century, Asia emerged as the fastest-growing economy, outperforming all other regions. This was in stark contrast to the previous four-and-a-half centuries when Asia had stagnated whilst other regions progressed. For instance, in 1500, Asia accounted for 65 per cent of the world's GDP, which reduced to 18.5 per cent in 1950. Since 1950, however, the Asian share has doubled.[5] Many view this increase in Asia's relative contribution to world GDP – compared to the West – as the 'recovery of Asia', or the transition of power from the West to the East.[6] The magnitude and speed of economic growth in maritime Asian countries has brought the Asia-Pacific region into prominence – providing one of the intersecting interests of major maritime trading and strategic powers, which eventually go on to shape global events. The coinage, 'Indo-Pacific', is the contemporary reference to this landscape.[7] The 'Indo-Pacific' usually refers to the geographic region bound strategically by India in the west, the US in the east, Japan in the north, and Australia in the south. *(It is no coincidence that this forms the 'Quad'.)* This region encompasses over 51 per cent of the earth's surface and includes approximately half the world's population. The UN has predicted that by 2050, seven out of ten people will live in the Indo-Pacific region.

Why Should the Rise of the Indo-Pacific be of Interest to the World?

During the June 2022 edition of the Posidonia shipping exhibition – considered to be one of the biggest international events of maritime trade, held biannually in the Greek capital, Athens – it emerged that politics was becoming the principal driver of the shipping market, surpassing economics. Whilst the large bulk of the debates alluded to the Ukraine-Russia conflict and Iran, this resonates with the centuries-old aphorism attributed to Pericles, the famous Athenian General. About 2,500 years ago, Pericles is believed to have said, *"Just because you do not take an interest in politics, doesn't mean politics won't take an interest in you."*[8] In the present, intricately connected world order, geopolitical development in any corner of the world almost instantly affects the entire globe. The rise of the Indo-Pacific is one of the major geopolitical developments of our time and must, therefore, be of serious interest to policymakers around the world.

As the rise of the Indo-Pacific increasingly influences geopolitical trends, some concurrent developments – including the speed and extent of disruptions caused by technologies of Industry 4.0; the impact of the COVID-19 pandemic, especially on the disruption of global logistics supply chains; and some of the ongoing conflicts around the world – could have a cumulative effect that could potentially alter the course of history. A review of history reveals that the significance of many of these events was recognised only once they unfolded. Viewed with the benefit of hindsight, it is clear that at least a few of these events did, indeed, alter the course of world history. Three examples stand out.

- First, the conquest of Constantinople by Sultan Mehmed II of the Ottoman Empire, on 29 May 1453, marked the end of the sprawling Byzantine Empire. In the period that followed, this event influenced the discovery of new sea routes, aided in the emergence of Europe as the global leader, and subsequently led to the establishment of the primacy of European sea power.
- Second, the 1648 Treaty of Westphalia, which marked an end to the Thirty Years' War in Europe and established lasting peace in the region, established the concept of nation States, separated religion from State, and consolidated the modern world economy and a Eurocentric anarchic world order.
- Third, the Industrial Revolution of the 1760s, which is also referred to by many scholars as the '1st IR' or 'Industry 1.0', was a hugely important event as it heralded a fundamental shift from agricultural societies to industrial ones. Of even greater import, it eventually led to the conception of world economy cycles – also called Kondratiev Waves – based upon global cycles of technological innovation. Each of these economic cycles ushered a new set of transformations driven by technology. Thus, the Industrial Revolution fathered the Machine Age, the Information Age, and now, Industry 4.0 (which is the confluence of technologies such as Artificial Intelligence [AI], Autonomous Vehicles, 3D Printing, Blockchain, the Internet of Things [IoT], Big Data analytics, Machine Learning [ML], etc.). Industry 4.0 is powering applications that are changing business processes across the services sector as well as the manufacturing one. Amongst these are the foundational pillars of the shipbuilding industry.

NOTES

1   A term used in many studies to refer to the latest applications that are built on digital technologies like Artificial Intelligence (AI), the Internet of Things (IoT), 3D Printing, etc.

2   Daniel Todd and Michael Lindberg, *Navies and Shipbuilding Industries: The Strained Symbiosis*, (Westport: Praeger Publishers, 1996), 10–11.

3   BS Randhawa, "Indian Shipbuilding: Key to Maritime and Economic Security," *Indian Defence Review*, vol 25, No 1 (2010, January-March, http://www.indiandefencereview. com/spotlights/indian-shipbuilding-key-to-maritime-and-economic-security/

4   Paul Kennedy, *The Rise and Fall of Great Powers*, (Great Britain: Fontana Press, 1989), 480-95.

5   Angus Maddison, *The World Economy: A Millennial Perspective*, (Paris: OECD Publishing, 2001), 142.

6   Joseph S Nye, Jr., "Understanding 21st Century Power Shifts", *The European Financial Review*, 24 June 2011, http://www.europeanfinancialreview.com/?p=2743, accessed 11 April 2018.

7   Indrani Bagchi, "Peaceful Periphery on the Seas - Why 'Indo-Pacific' is Nudging 'Asia-Pacific' off the Table Even as China's Shadow Looms," *Times of India*, 29 May 2018.

8   Lloyd's List, "Pericles Was Right; Shipping Can't Ignore Politics," June 10, 2022, https://lloydslist.maritimeintelligence.informa.com/LL1141191/Pericles-was-right-Shipping-cant-ignore-politics.

2

Shipbuilding: A *Tour de Horizon*

Introduction

The shipbuilding industry encompasses the design, construction, and maintenance of ships, crafts, boats, offshore platforms, etc. It has evolved over centuries – from the early days of boats like paddling and sailing canoes, galleys with oars, wooden sail ships, and engine-propelled steel ships, to the modern-day autonomous unmanned vessels. The primacy of ships[1] and boats in human social construct has been pivotal, as it has been a popular medium for transportation and trade.[2] Humans are believed to have been using various types of boats even before the wheel was invented.[3] Considering that nearly 80 per cent of the world's population lives within sixty miles of coastline and two-thirds of the global economy is derived from activities that involve the seas in some form,[4] the relationship between human evolution and the seas is symbiotic. Shipbuilding has been the foundation of this relationship.

Shipbuilding has therefore been instrumental in the overall growth of societies, involving the multi-trade industrial complex that employs large numbers of people of different specialties. In the era of wooden ships with sails, shipbuilding involved designers, carpenters, dubbers, caulkers, joiners, riggers, and other craftsmen.[5] Over the centuries, the use of iron, new design techniques, construction processes, modern machinery, and electronics and data processing systems added to the complexity of shipbuilding.[6] Shipbuilding was the earliest vehicle and facilitator of globalisation, and it continues to be a principal component of the modern world system.[7] This is underlined by the fact that even today, any product has a ninety per cent chance of having travelled in part or whole on a ship, for at least some part of its journey to the consumer. Just as merchant ships form the backbone of trade and commerce, warships and submarines form the basic building blocks of sea power[8] – an extension of military power on the seas.

To begin, it is important to examine how shipbuilding contributed to nation-building and emerged as a key element of the modern world economy, with sea power becoming a significant component of comprehensive national power.

Interplay of History and Strategy: Rise of Sea Power

Around the tenth century AD, there was a substantial increase in inter-regional trade. Commodities exchanged at the inter-continental level were now transported by sea rather than over land. Between the tenth and fifteenth centuries, Venice emerged as the European trading hub for intra-Mediterranean maritime trade and inter-continental trade through the Classical Silk Routes. Chinese products were traded via caravan routes to ports in the Black Sea, while Indian and other Asian products were transported via Syria and Alexandria. Major ports were developed in the Red Sea under the Ayyubid rulers (1170-1260 AD) and the Mamluks (1250-1517 AD) in Egypt.[9] However, in the latter half of the fifteenth century, the established trade link between Europe and Asia was cut off – a result of the cumulative effect of developments, including the fall of Constantinople, the rise of the Ottoman Empire, the collapse of the Crusader states in the Levant, and the end of the Mamluk regime in Egypt. The Europeans, therefore, made concerted efforts to venture seawards and establish trade routes to Asian destinations.

The early voyages undertaken by Columbus, Vasco da Gama, and Magellan in sixteenth-century Europe – referred to by many as the 'expeditions of trading and plundering'[10] – transformed the world's oceans from 'impassable barriers' into what Alfred Thayer Mahan called the 'great commons'. These voyages were made possible because of the development of ships capable of withstanding the rigours of technology, stresses, and strains of ocean voyages; new means of accurate navigation on the open seas; and shore-based infrastructure capable of sustaining the new merchant fleets and the navies that protect them.[11] Although the seas and oceans evolved as the 'global commons' facilitating trade, commerce, and connectivity, commercial rivalry also increased.

The breach of the ocean barrier led to a fundamental change in global trade patterns. In the early Middle Ages, trade was largely localised and only luxury goods were exchanged over long distances. There was no exchange of "bulk" goods of "staples" across intermediately sized areas, hence, there was no production for such markets.[12] Oceanic trade now flooded the European ports with not only gold, silver, precious metals, and spices, but also 'staples' in 'bulk' – like seafood, whale oil, seal oil, sugar, indigo, tobacco, rice, fur, timber, new plants like potato and maize, to name a few. Thus emerged the first-ever world economy system and a globalised world order.[13] Capitalism, which had initially been limited to the appropriation of surplus value by landowners, now expanded to include the appropriation of surplus of a country by another country that was more powerful and capable of enforcing an unequal exchange.

Sea power became a significant component of comprehensive national power and was applied to influence political, military, and economic decisions on the world's oceans.[14] Merchant ships, warships, auxiliaries, maritime infrastructure, bases, and their respective personnel, encompassed sea power that was now used for military, diplomatic, constabulary, and benign roles to serve national interests. However, the underlying principle of the evolution of sea power was to protect and expand sea trade and commerce which, in turn, enhanced sea power itself.[15] As global maritime trade and commerce grew and became a major source of nations' wealth accumulation, conflicts of interest in sea commerce often led to wars. Even the conflicts that emerged for other reasons began getting influenced by the desire to control the seas. As sea power became a significant component of the comprehensive national power, the shipbuilding capability of a nation became a significant benchmark for aspiring great powers. Major sea powers were often leading shipbuilding nations as well.

As global sea trade and commerce grew, the demand for larger ships increased. Countries preferred to use their own vessels for their shipping business, wanting to ensure the safety and security of their ships during voyages and whilst discharging cargo at foreign ports. A peaceful shipping business thus became key to the assured wealth accumulation of a country. A powerful navy – with a long reach – could be used against pirates and raiders during peace, as also against an adversary during war.[16] However, maritime trade exhibited two distinct regional characteristic patterns: one in the Indian-Pacific Ocean region, extending from the east coast of Africa to the west coast of the Americas, whilst the other lies in the Mediterranean Sea.

In the Indian Ocean, the sea power of countries and empires varied from time to time, depending on the strategic considerations and the wealth that rulers accrued from seaborne trade. The Chinese, Indonesians, the peoples from the Southern Indian kingdoms of Vijayanagar and Kalinga, Cholas, Arabs, Persians, and Sultans of Muscat – each group has developed formidable sea power in different periods of history. However, no nation or empire ever managed to dominate the Indian Ocean as a whole. No political power attempted to control the sea-lanes and the long-distance trade of Asia, except in localised regions like the Persian Gulf or the inland sea of the Indonesian islands. The commercial contracts included established conventions, and these were enforced by means of collective sanctions against any defaulting member.[17]

During the medieval period, two primary trade routes were prevalent in the Indian Ocean. The first route connected commercial cities in the Red Sea and the Persian Gulf which had been united by the common bond of Islam post the seventh century. The second route was the long trans-oceanic route to India, the Indonesian archipelago, and China. The latter trade circuit, which was bound by the Pacific at one end and extended to the Mediterranean at the other, constituted the foundation of the pre-Columbian world economy in both the East and the West. Highly skilled professional merchants managed the commercial traffic which flourished as a result of the high degree of political freedom and neutrality of port cities.[18]

In the Mediterranean, however, dominant powers exercised control over the vital sea-routes to control both economic resources and political settlements, from Graeco-Roman times and perhaps even earlier periods of history. One of the key components of the ancient Greeks, Phoenicians and Romans' sea power was the Trireme warship – triple-deck Galleys with oars.

During the fourteenth and fifteenth centuries, Italian city-states like Venice, Florence, and Milan, among others, emerged as power centres; the institutional basis of world trade in the Mediterranean region underwent a new development. The commercial rivalry between Genoa and Venice erupted into open naval conflicts, and the Venetian encounters between the Christian and Muslim fleets fused together the interests of the merchants and states. The Italian experience was reproduced later in Seville, Lisbon, Amsterdam, and London. As the voyages of the sixteenth century established alternate trade routes through the vast oceans, the volume of sea commerce increased. As a result of this, two distinct regional trading systems of Europe and Asia were bridged, thereby transplanting seedlings of the Mediterranean pattern in the Indian Ocean. This was made possible once the Iberians developed long-distance armed merchant shipping, floating fortresses, and warehouses, making it possible to extend the area of oceanic control from home bases in Europe and establish new bases in remote places in Asia. Advances in shipbuilding technology were an important factor.[19]

By the first decade of the sixteenth century, therefore, the period of peaceful sailing in the Indian Ocean was over. Instead, the way had been paved for an era where whoever controlled the sea was in the position of overwhelming commercial and political superiority.[20] As the Portuguese, Dutch, and British merchants – with their armed shipping – ventured into the Indian Ocean and beyond for trade, they sought to acquire 'stations' at the far end of the trade route by force or favour. These 'stations' ensured safe handling and exchange of cargo, smooth functioning of agents, safety of the crew, and maintenance support for the ships. They promoted trade and eventually grew into colonies, becoming a significant part of the history of the medieval and pre-modern world. During the early colonisation era, there were pirates and raiders who posed dangers to the safety of the shipping routes; these were addressed by creating 'stations' along the route, such as the Cape of Good Hope; St. Helena; Mauritius; Gibraltar; Malta; Louisburg; the Gulf of St. Lawrence, etc., which were important not for trade but strategical value for defence and war.[21]

As Great Britain began consolidating its colonies in the Indian Ocean region, it reinforced the road to India by acquiring St. Helena, the Cape of Good Hope, and Mauritius. Once steam-propelled ships made the Red Sea and the Mediterranean practicable, Britain acquired Aden and later established itself at Socotra. These acquisitions were either by *ab initio* domination of the indigenous peoples or by defeating the earlier established European powers, such as the Portuguese, Dutch, or French. Sea power made Great Britain rich, in turn, protecting the trade that generated its wealth.[22] Great Britain also used sea power to generate revenue through fee collection in return for protecting neutral countries' ships in transit, through the means of a convoy escorted by British warships, thereby mandating it under the Convoy Act of 1798.[23] Great Britain also used sea power to deny sea trade to the French, effectively checking the rise of the French Empire after the French Revolution.[24]

In the early nineteenth century, Britain emerged as the dominant sea power of the world and maintained its position well into the twentieth century. The United States emerged as one of the leading sea powers by the end of World War I. In the 1930s, US shipbuilding surpassed that of Britain and other countries, making it the leading shipbuilding nation. After World War II, manufacturing bases shifted to emerging Asian markets including Japan, South Korea, and later China, thereby crowning them as the world's leading shipbuilding nations. Warship production along with weapon systems development, however, was kept under state control, and Western countries maintained dominant sea power status, resulting in a situation where major shipbuilding nations were not necessarily great sea powers.

Sea power became central to the strategic calculus of great powers after the voyages of the sixteenth century, but long sea voyages were sustained because of the increase in volumes of international trade and the capitalist-based modern world economy. There was also an increase in the relative productive capacity of Europe compared to the rest of the world. This was a result of some fundamental changes in the mode of labour control in Europe, which effectively ended feudalism and laid the foundations of the modern world economy. The increase in productive capacity enhanced the shipbuilding capabilities and capacity of European countries, thus enabling seafarers to venture into the deep seas.

Interplay of Economics and Strategy: The Modern World System

At the end of the fifteenth century and the beginning of the sixteenth century, there was a fundamental transformation in the mode of labour control in England. Workers were hired and wages were paid based on their skills and specialisation. This was different from the erstwhile feudal system, where a worker worked as captive labour for the feudal lord. The system of tenancy and wage labour favoured specialisation and higher skill levels. This transformation started with agriculture and quickly diversified to textiles, shipbuilding, and metal wares in northwestern Europe. Advancements in structured finance and commercial systems ushered in the capitalist-based economic system. The principal motivation of this system, however, was the landowner's appropriation of the labourer's surplus-value; by extension, the stronger country's appropriation of the weaker country's surplus-value by maintaining and enforcing unequal exchange. The State provided a strong military with a significant naval component, shipping technologies, and an advanced bureaucracy to further business interests. Whenever required, the State intervened in support of enterprises that were domiciled in their country.[25]

Europe emerged as the centre of the capitalist-based economy, with a three-tiered structure, which soon evolved into the modern world economy system. The building blocks of this system were individual territorial entities which were culturally homogeneous, had a structured bureaucracy, raised a standing army and a powerful navy, and had a diverse economy. This European world economy was firmly established by 1557 and interacted with the other contemporary world systems like the Ottoman and Russian empires and the Indian Ocean proto-world economy. North-western Europe emerged as the 'core' area leading in agriculture, textiles, shipbuilding, and metal wares. Mediterranean Europe, including Spain and Italy, became a 'semi-periphery' area, specialising in high-cost industrial products like silk. These countries also implemented advanced credit and finance systems like the core region, enabling sharecropping[26] and limiting exports to other areas. North-eastern Europe and the Western hemisphere, including Iberian America, became peripheral areas where investors and enterprises domiciled in the core areas could use slaves – often transported by sea – and coerced-cash-crop-labourers to grow and export grains, bullion, wood, cotton, and sugar to the core areas and other markets as per demand.

The Eurocentric world economy had stabilised by the 1640s, with core nations having strong state-mechanisms to meet the needs of the capitalist landowners and their merchant allies. This contrasted with the weak state-mechanisms in peripheral nations, which would be liable to the enforcement of unequal exchange, and thus render them more vulnerable to interventions through war, subversion, or coercive diplomacy. Sea trade was the backbone of the modern world system and sea power emerged as the principal instrument of global dominance. Shipbuilding, therefore, became a major component of the military industrial complex of an emerging great power. The stability and successful existence of the three-tiered, capitalist-based modern world system of unequal distribution of rewards has been attributed to three factors. First, the core countries with major wealth accumulation became dominant world powers with the highest concentration of military strength. Second, the advantaged sections worked to preserve the system since they related their own well-being to the survivability of the system. Third, the core was insulated from the larger lower stratum of the periphery by a smaller middle stratum of a class – the latter simultaneously the exploiter and the exploited, as a semi-periphery entity.

The system of the 'State' was the building block of the Eurocentric modern world economy, formalised during the Westphalian peace negotiations at the end of the Thirty Years' War that waged between 1618 and 1648.[27] As a part of these negotiations, a common provision in the treaties at Münster and Osnabrück resulted in the creation of about 300 sovereign political entities which had complete independence; they were recognised by the collective decision of the congress. Westphalia, therefore, marked a historical watershed in European affairs, wherein there was recognition of the existence of a nascent society of independent States, and any new State being formed was to be recognised by this international body. These sovereign political entities were no longer required to derive legitimacy from the Pope, thus rendering governance independent of religion. Although formally, the principles applied only to Germany, they spread to the entirety of Europe and carried a corollary of the principle of 'non-interference' by one State in the internal affairs of another. Westphalia thus marked the formal birth of the modern nation-state system that is characterised by multiple sovereign entities in an anarchic setup.[28] Another salient yet one of the most significant fallouts of the Thirty Years' War was concerning the interests of the Dutch that prevailed on Habsburg. Whilst it signalled an end to the attempts to regain supremacy of the Hapsburg Empire, it ushered in an era wherein the dominant world power had to be a dominant sea power. This trend remained consistent in the following centuries when the British and the US – at a later stage – emerged as dominant powers.

There was a system-wide recession from 1650 to 1730, following which began the Industrial Revolution in 1760; England emerged as a dominant power, surpassing the Netherlands and later France as the core. There was a transition from a predominantly agricultural economy to an industrial-based economy. The countries in the core, with Great Britain in the lead, came to possess ever stronger military capability, with sea power primacy which extended their geographical reach. All other world systems collapsed and were integrated into the new global order. Russia entered semi-peripherally; Latin America, Asia, and Africa were absorbed in the periphery. Japan, because of a combination of its state machinery and remoteness from the core, could graduate quickly to semi-periphery. The core consolidated its position further by taking the lead in technological innovations. In the ensuing period, the rise of great powers in the anarchic world order became increasingly correlated with technological innovations that fuelled the cycles of the world economy, whilst the centrality of sea power continued to be a necessary condition.

Interplay of Economics and Technology: The 'K Waves'

A system of institutionalised and structured banking, credits, and insurance was well-established in Europe because of the Financial Revolution. Adequate finances became available in the form of long-term low-interest loans for scientific research and development. A series of innovations ushered the Age of Machines in the 1760s; moreover, there was an unprecedented drive to increase productivity in north-western Europe, with Britain in the lead.

As the Industrial Revolution led to the increased production of goods, there was an increase in the volumes of trade and commerce, a greater accumulation of wealth, and an improvement in the quality of life. However, in the ensuing period, the world economy experienced long-term cyclic patterns of growth and recessions, and this has been a subject of study by many global economists. In 1939, Joseph Schumpeter – an Austrian economist from Harvard University – concluded that since 1780, there have been several extended periods of extraordinary economic growth, followed by periods of reduced growth – punctuated by deep recessions. A rough periodicity of 50 to 60 years was also observed, however, it varied from country to country. Schumpeter named these cycles 'K Waves' after Russian economist Nikolai Kondratiev; the latter was credited with identifying three waves of economic development in the industrial era – from 1790 to 1920 – and had projected the Great Depression of the 1930s.

Schumpeter's theory considers technological innovation as the main driver of economic development and argues that the driving force of change is the "perennial gale of creative destruction".[29] It is a process that begins with innovative action, continues with imitative action, and ends with the destruction of the old economic structure. Innovation is driven by entrepreneurship and efforts to break the repetitive creation of existing products and processes. A significant characteristic of each 'K Wave' was that the country with the maximum technological innovations emerged as the leader of the global economy and evolved as a major sea power to dominate the anarchic world order. Sea power facilitated sea commerce and the shipbuilding industry, and each 'K Wave' significantly transformed shipbuilding technology.

The first 'K Wave' – which occurred from the 1780s to the 1840s – is sometimes referred to as the Age of Machines, driven by the invention of the steam engine.[30] Since cotton and wrought iron were available in large quantities and at reduced prices, there was a boost in textiles and structural engineering. Coal-based steam engines and machine tools replaced water, horse, and wind as sources of power. As the application of steam engines became more widespread, there was great demand to transport coal over large distances, and major projects interlinking rivers through canals came up. Subsequently, advancements in steam engines and iron-making facilitated innovative applications of technology and ushered the railways as an alternative to canals for heavy goods transportation.

In the shipbuilding industry, initially, iron armour was used on wooden hulls which then graduated to iron-clad ships. There was also a transition in propulsion, from paddles to screw propellers. This was followed by the introduction of surface condensers, and finally, the steam engine for the main propulsion. However, during the period of the first 'K Wave', the proliferation of iron hulls in shipbuilding was slow, and the majority of the fleet strength continued to be that of the wooden-hulled ships – in both the applications of the merchant marine as well as the warships – though there was an exponential rise in the numbers and sizes of these ships along with their contemporary designs.

The emergence of the railways triggered the second wave of technological transformation, providing the impetus for a major expansion in iron production; the adoption of more efficient smelting technology, like the hot blast; and the search for better ways of making steel. There was an increase in the use of iron in shipbuilding, and iron ships later paved the way for steel ships. Advancements in the steam engine, from single-expansion engines to two- or three-stage compound engines with high-pressure superheated steam, were first proven in the cotton mills and subsequently found extensive application in ships.

The opening of the Suez Canal in 1869 hastened the decline of sail ships, since prevailing winds in the Mediterranean, blowing from west to east, mandated that sail ships go around the Cape on their homebound journey from India and China. Large ships were introduced as liners, long-haul cargo streamers, and oil tankers.

During the third 'K Wave', iron was replaced with steel and petroleum and electric power industries emerged. The internal combustion engine was developed. The fourth 'K Wave' started in the 1940s, driven by the development of petrochemicals and the expansion of the auto industry. In the 1960s, the computer or digital revolution led to the development of semiconductors, mainframe computing (1960s), personal computers (1970s and 80s), and the internet (1990s). Table 2.1 summarises the period of each 'K Wave', the cluster of technologies, and the changes to shipbuilding during these waves.

Table 2.1: 'K Waves' and Technology Lead Sectors

Ser	'K Wave'	Time-period	Lead sectors	Changes to Shipbuilding
1	First	1780s to 1840s	Steam power, cotton textiles and iron	Iron used for shipbuilding, the changeover from paddle to screw propulsion, surface condensers and steam engine for propulsion[31]
2	Second	1840s to 1890s	Railways, iron, and steel	Increase in the number of Iron-armoured and then iron-clad ships. Use of steel instead of iron. Compound steam engines with superheated steam. Liners and long-haul cargo streamers and oil tankers.
3	Third	1890s to 1940s	Electricity, chemicals, and automobiles	Introduction of steam turbines, destroyers, dreadnoughts, diesel engines, rivets to welding, wireless telegraphs, radars, submarines, aircraft launch from ship
4	Fourth	1940s to 1980s	Electronics, synthetic materials, petrochemicals and nuclear	Diesel marine engines, combined steam and gas turbines, nuclear propulsion
5	Fifth	1980s to 2015/ 2025	IT and aerospace	Stealth, network centricity, communications, autonomy

As summarised in Table 2.1, the Industrial Revolution completely transformed shipbuilding technology. This was, in part, because sea trade was the backbone of the modern world system and hence, the advancements in technology were applied to build ships that could support the world's rising trade. Since sea power formed a major component of the comprehensive national power, new technologies were used to build more powerful warships. The narrative thus establishes how the interplay of strategy, economics, and technology influenced shipbuilding during this period of history. The multi-dimensional model of analysis will now be used to sketch in broad brushstrokes the evolution of shipbuilding under the comprehensive influence of the parameters of history, economics, technology, and strategy during four broad periods when there were major transformations in shipbuilding. The first period covers history up to the fall of Constantinople, which was a turning point since trade connectivity between Asia and Europe through the traditional Silk Route was broken. The periods that follow are the era of the Great Voyages (1500 to 1750), the Age of the Industrial Revolution (1760 to 1860), and the period of the World Wars (1870s to 1940s), since each of them transformed shipbuilding.

History up to the Fifteenth Century AD – Until the Fall of Constantinople

Since early history, humans used the seas and oceans for their livelihood, trade, and adventure. Therefore, boat- and ship-making improved continually, with the experience of seafarers and the demands of society. There were unique regional characteristics and flavours in ship design to cater to the local conditions and demands of sea commerce.

The types of ships used in the Indian Ocean region can be broadly divided into three groups. In the western half as far as Bengal, there was an Indo-Islamic tradition of shipbuilding which produced several common hull shapes with vessels rigged with lateen sail. In the Indonesian islands, Malaya, and Burma, fast and light boats like the Prahu and Sampan were predominant. Although they were ocean-going vessels, they were restricted in cargo capacity and were used in inland seas. The third type was the Chinese Junk – the carrier of choice in the Indonesian archipelago and the Far East – which were massive rectangular barges also known as "castle ships" with multiple floors fortified with weapons.[32] The geographical frontiers separating these generic ships were as marked as the ethnic and linguistic boundaries in the Indian Ocean.

Indians designed vessels, ranging from small boats for river and coastal use to large ships that had double masts for long sea voyages. These large ships were stitched together with rope rather than being held together by nails.[33] Although they were suitable for local trading requirements, they were not sturdy enough to withstand the typhoons of the China Sea, and not adequately armed to deal with pirate activity off the China coast.[34] During the period of Islamic expansion towards India and the Indonesian archipelago prior to the twelfth century, the shipbuilders in the Persian Gulf and the Red Sea upgraded their undecked smaller crafts into bigger vessels with fuller decks, offering a more watertight construction.

Minoans based in Crete (3000 BC to 1500 BC) used sailing ships with oars and are believed to have invented the keel.[35] Egyptians (3200 BC) built ships from papyrus and later used cedar planks joined at the ends. The ships had no keel, just a rounded hull under the waterline, rendering them susceptible to drift. The sea-going ships were held together by means of a rope-spine or a hogging-truss. Phoenician ships had a keel plank, to which a garboard strake was attached by means of mortise-and-tenon joints, all of which were pegged.[36]

Shipbuilding in Mediterranean Europe evolved from science and technology that were rooted in the works of ancient Greece and Rome. The Roman Empire was split into the Western and the Eastern Roman empires in 330 AD, with the capital of the Eastern Roman Empire in Byzantine – later named Constantinople. Whilst the Western Roman Empire disintegrated by 476 AD, the Eastern Roman Empire – referred to as the Byzantine Empire – lasted for another thousand years.[37] The scholars in the Byzantine Empire interpreted and advanced Greek literary, philosophic, and scientific works.[38] There was also a significant outflow of scientific ideas from the Arabian to the Byzantine territories through translations, flowing from Arabic to Greek.[39]

In Europe, the city-states facilitated shipping. In Venice – which was a dominant player in shipbuilding – the State facilitated shipping by being the largest shipbuilder and leasing the state-owned galleys to the merchants. Arsenal – a public shipyard created in 1104 – was its largest enterprise. Unlike ancient Roman shipbuilding where initially the ship's hull was constructed and held together with watertight cabinetwork of mortice and tenon, with ribs and braces inserted in the second stage, in eleventh century Venice, the keel and ribs were made first and a hull of nailed planks was added, using fibre and pitch to make the ships watertight. Later, the stern-post rudder replaced the trailing oars as a more effective means for steering; its power was strengthened by the use of cranks and pulleys.

There were improvements in sails, notably the introduction of the triangular lateen rig set at an angle to the mast, instead of a rectangular sail square of the mast. There was a long-run increase in the size of the ships. In 1270, the compass was introduced, and along with improved charts, it facilitated year-round sailing in the Mediterranean. Whilst general-purpose cargo ships called 'cogs' were built by private shipyards, galleys – meant for passengers; high-value cargo vessels; and naval ships were built by the Arsenal.[40] Galleys were light, narrow, fast, and manoeuvrable and were far superior warships, with speed and consistency unmatched by any rival.[41]

The unprecedented ascension of Venice was particularly driven by its maritime expansion as a result of improved shipbuilding techniques; the use of the compass and other improvements in navigation,[42] and the development of institutional banking, finance, credit solvency, and diplomatic service. The introduction of the compass for navigation and the sunglass for measuring time at sea made it possible for ships to navigate in bad weather and make two return journeys a year from Venice to Alexandria instead of one.[43]

Chinese ships were much bigger than their European counterparts, more seaworthy and comfortable, with watertight compartments, several cabins, and a capacity to navigate over long distances – as far as Africa. The treasure ships had nine masts, and smaller ships had multiple masts. Big ships had fifteen or more watertight compartments, so a partially damaged ship would not sink and could be repaired at sea. Under the third emperor Yung-lo of the Ming Dynasty (1368-1644), a fleet of 62 ships carrying 28000 personnel sailed from Fukien to Calicut. Under Zheng He (1404 -1433), subsequent voyages were undertaken, with claims to have reached as far as Hormuz, Aden, and the east coast of Africa.

The Asian merchants were familiar with the seasonal wind patterns and problems of the Indian Ocean. They had experienced pilots and because of scientific works on astronomy and navigation, their navigational instruments were comparable to those of the Europeans.[44] As Atlantic ships began to arrive in the Indian Ocean at the end of the fourteenth century, Asian shipbuilders modified their hull design to produce stronger versions of the traditional craft, which were identical models of Atlantic ships.[45]

During the early fifteenth century, there were repeated attacks by Turks on the Byzantine Empire, which caused fear and instability in Constantinople and triggered the large-scale migration of scholars to Italy and other destinations in Europe. These scholars fuelled scientific resurgence and the Renaissance in Europe, initiating many technological innovations in shipbuilding. The fall of Constantinople in 1453 cut off the traditional Silk Route from Asian destinations and exerted pressure on the Europeans to venture seawards and establish alternate routes to trade with Asian destinations.[46]

1500 – 1750 AD: Era of the Great Voyages

In the latter part of the fifteenth century, ambitious interaction between Europe and the rest of the world began, with Portugal in the lead. Vasco da Gama, Magellan, and others ventured into the oceans, establishing sea routes to Asia and the Americas. These successful voyages were a result of the state-sponsored Atlantic exploration, research on navigation technology, training of pilots, and documentation of maritime experience in the form of route maps with compass bearings and cartography. The deep-sea fishermen from Portugal provided insights into Atlantic winds, weather, and tides. Once new routes were found, other European countries followed suit.

There were two basic types of ships in Europe – 'galley' and 'round ship' – that served as models of development on which incremental design modifications were done from the sixteenth century till the introduction of steam-powered ships.[47] Portuguese shipyards in Lisbon and Oporto adapted the design of their ships – 'caravels'[48] – to the Atlantic sailing conditions to facilitate deeper penetration into the south Atlantic, and later, for much longer routes around the cape. Portugal took the lead in opening European trade, navigation, and settlement in the Atlantic islands, as well as developing trade routes around Africa, into the Indian Ocean, to China and Japan. It emerged as a major shipper of spices to Europe, usurping the role from Venice. It transferred the cane sugar production and processing technology to the islands of Madeira and São Tomé, and later to Brazil. It inaugurated the slave trade to provide a labour force for the industry and carried about half the slaves who were shipped to the Americas from Africa between 1500 and 1870.[49]

In Asia, trade was largely conducted by merchant communities and the rulers of the kingdom of Vijaya in south India. Later, the Mughals derived income primarily from land taxes and had no significant financial interest in trade activities. China withdrew from an active role in Asian trade in the fifteenth century and imposed tight controls on private trade, and an embargo on trade with Japan. Europeans gradually increased their presence and began to dominate by setting up remote bases. The Portuguese established a base in Macao in the 1550s and participated in clandestine trade with China and Japan.

The Dutch later surpassed Portuguese dominance as it rose to a commanding position in trade and shipping. The Dutch per capita was the highest in Europe from 1400 to 1700 and peaked between 1600 and the 1820s,[50] as it dominated shipping and trade in the Baltic, surpassing the German traders' lead position. This was facilitated by the establishment of new sea routes for trade in the north which, though longer, provided a cheaper mode – especially for heavy cargo. The State facilitated the development and mass production of cheap general-purpose cargo vessels called 'fluyt' and large 'factory ships' – equipped to catch, preserve, and package Herring fish at sea, thus establishing them as leaders in the Herring trade. Later, the British government financed and encouraged research into astronomy, terrestrial magnetism, and the production of the first reliable chronometer and nautical almanacks. They also demonstrated the efficacy of sauerkraut and citrus juice in preventing scurvy.

The Dutch employed labour-saving machinery such as the wind-driven sawmill and the great cranes – they lifted and moved heavy timbers and placed masts on ships, reducing the cost of ship construction in Holland by one-third/half the cost of a similar ship in England. In the early seventeenth century, the distinction between a warship and a merchant ship was not clear-cut and exclusive, and most large merchant ships were fortified and well-armed to defend themselves if required. The merchant fleet secured national trade and was a ready source of large ships' supply that could be improvised by the navy.

In the sixteenth century, shipbuilders in Asia adopted some techniques, like the use of lateen rigging and iron nails as plank fasteners; these techniques were prevalent in the Mediterranean. Teak-built ships constructed in the naval yard of Goa were strong and regularly sailed on Atlantic voyages to Portugal.[51] By the mid-seventeenth century, differences in western and the eastern ship designs had considerably reduced, with the Asian shipbuilders having adopted Western techniques – especially with respect to improved defence and larger cargo-carrying capacity.

The latter part of the seventeenth century marked the rise of highly professional navies and prolonged campaigns at sea. There was a gradual transition in the technique of warfighting, from boarding and hand-to-hand combat, to the reliance on guns. As weapons began to be recognised as the major factor in naval victory, structural alterations were made in the fighting ships to accommodate more and bigger guns. The British Navy took the lead in this adaptation with the construction of distinctive fighting types: the great ship and the frigate.[52] Larger British merchant ships usually had designs like warships, with the greatest width at three to five feet above the waterline at the gun-deck and the lines of the sides tapered sharply to the keel. This was different from the flat-bottom design of the Dutch ships. Though it resulted in better sailing quality, it narrowed the hold, reducing cargo-carrying capacity.

The Industrial Revolution in Europe helped to expand their presence and consolidate domination, wherein the impact was more pronounced since an integrated world economy system was already in place.

1760 – 1860 AD: Era of the Industrial Revolution

Significant technological developments around the 1760s ushered in the 'Age of Machines'. These developments occurred in Europe, further strengthening European dominance in a world that was already integrated through the extensive network of sea trade. A major technological transformation was the use of coal and steam for propulsion, and iron for tooling. Steam propulsion replaced the erstwhile sail and oars in ships. Wooden hulls began to be replaced with iron-cladded, and later, steel ships, making them far superior in strength and design. The weapons and armament technologies also added to the differential between the European powers and the Asian powers – largely isolated and inward-looking during this period.

In the initial period, iron armour on wooden hulls began to be used, which then graduated to iron-clad ships. There was also a transition – from paddles to the use of screw propellers for propulsion between 1843 and 1845. This was followed by the introduction of surface condensers and finally, the steam engine for main propulsion, though the proliferation was slow. A vast majority of the fleet strength continued to be that of the wooden-hulled ships, though there was an exponential rise in the numbers and sizes of the contemporary design.

The first steam ship *Comment* was launched in Britain in 1812. The development of steam engines required a stronger hull, and therefore, wooden ships were gradually replaced by iron hulls. By 1850, the British had considerably improved the hull design, with wave line hull forms. The shipyards were constructing ships with a combination of iron frames, timber planking and copper sheathing. By the 1860s, British shipbuilders were constructing the largest number of ships in the world.

After the 1840s, there was an increase in the use of iron in shipbuilding that later paved the way for steel ships. Once steam engines began to be used extensively on ships and the Suez Canal was opened, large ships were introduced as liners, long-haul cargo streamers, and oil tankers. There was a significant change in the pattern of goods exchanged through trade. Whilst in earlier periods, the direction of flow was defined by the local needs of essentials and luxuries, this was not the case after the Industrial Revolution. Instead, raw materials from the colonies were sent to the industrial centres and finished products from the industrial centres were sent to the markets.

There also emerged a strong coupling between technology and industrial base on the one hand, and the military power and diplomatic influence of a nation-state on the other. Political diplomacy began to play a significant role in shaping the sphere of technology-economics-strategy. Competing powers reorganised their productive forces to respond with unprecedented surges in industrial production to support their war-fighting capabilities.[53] There was an unprecedented growth in the number of ships built, both for mercantile marine as standard ship designs and warships like destroyers, frigates etc., recommended for navies with global ambitions.

The 1870s – the 1940s: Period of the World Wars

Around the 1870s, iron began to be substituted with steel as an engineering material and as steam engines began to be commonly used for propulsion using screw-propellers, there was a shift from piston engines to turbines which were already proven. This resulted in higher efficiency and lesser weight, leading to higher carrying capacity. A variety of ships like coasters, grain ships, refrigeration ships, cattle ships, whale factory ships, and cable-laying ships became popular. In naval application, torpedo boat destroyers and Dreadnoughts were introduced. Diesel engines were also introduced for the main propulsion; wireless telegraphs made ship-to-shore communication possible; and radars were introduced. This period also saw the early application of submarines as well as the launch of aircraft from a ship with a flat-top deck design. During the war, to ensure shorter repair times, welding was used instead of rivets.

During the last two decades of the nineteenth century and the first two decades of the twentieth century, the British Royal Navy had formulated a framework of a model for a blue-water navy, based on a balanced fleet structure encompassing classes of standard types of warships.[54] This stimulated international demand and resulted in the growth of the shipbuilding industry from the 1870s till World War I, wherein British shipbuilders were in the lead. The steel, steam, engineering, and metal industries were well-established as ancillary industries.

Shipyards moved to those locations where iron and coal were available. In Great Britain, shipyards migrated from the south to the north and northeast. By 1882, Britain had 82 per cent of the world's shipbuilding market. Initially, steel was sparsely used for hull fabrication owing to its high cost; however, within a few years, new steel-making processes were developed and this resulted in the commercial availability of steel. By 1889, nearly 97 per cent of the tonnages were built with steel. In 1912, Britain started building ships with diesel engines. There were considerable improvements in mathematical modelling or design and tank-based ship model testing for resistance and stability.[55]

The standardised classes of ships prescribed for a model blue-water navy found worldwide acceptance by emerging powers and led to a surge in shipyards producing warships, especially in Great Britain and the United States. Dominating the force structure was the Battleship (BB). This was a capital ship, and was heavily armoured and armed, designed to engage the enemy's capital ships in major, all-out naval battles. The Royal Sovereign class of Britain served this role from the 1890s and was upgraded to the Dreadnoughts in 1906.[56] All the subsequent BBs adopted the upgraded design and were referred to by that name.[57]

The Cruiser (CC) occupied the second place in the hierarchy of class of ships, with two types – 'protected' and 'armoured' CCs. They first appeared in the 1870s. These ships had heavy guns, could do high speeds, and could be deployed for commerce raiding and scouting. The Torpedo boats were light, small, and fast craft, armed with several torpedoes and one or two small calibre guns. To counter the attacks, Torpedo-boat destroyers were designed. These evolved into Destroyers and subsequently were armed with torpedoes and more powerful guns and were designated as DDs. The final components were the naval auxiliary vessels which were the oilers and colliers at that time.[58]

Although the structure of a model blue-water navy originated from the British Royal Navy, in due course of time – Britain's Grand Fleet, Germany's High Seas Fleet, America's Great White Fleet, and Japan's Imperial Fleet – all imbibed this model. They not only served as the epitome of sea power, but also became symbols of their State's political and economic power. The standardised fleet structure infused demand for newly constructed warships worldwide. Shipyards then geared up to handle the special requirements of a warship's design and construction. This trend has lasted ever since, although the markets experienced large fluctuations in demand through the World Wars.[59]

Britain continued to dominate the world's shipbuilding market both in production capacity and use of advanced technologies, but encountered a major setback by the end of the First World War when orders of warships were cancelled, leading to over-capacity that further worsened during the 1930s' Great Economic Depression. Many shipyards were shut down and higher demands during the Second World War were met by shipyards which had been brought under government control. However, the British shipbuilding industry did not keep pace with the rest of the world in terms of currency of shipbuilding technology or in terms of augmenting the production capacity. The result was a steady decline in the position and market share.[60]

In the US, there was a transformation in the shipbuilding industry after the American Civil War (1861-65). Shipyards took to the production of iron-hulled and steam-powered ships. By 1880, scientific methods were being used extensively in shipbuilding. The American shipyards started working closely with iron and steel manufacturing factories, and many shipyards set up in the late 1800s and early 1900s were jointly owned by the iron and steel company. In the 1870s, Admiral Alfred Mahan's theory of sea power began to revive the US Navy. This led to the mastery of European-developed, advanced methods of shipbuilding. The development of shipbuilding was driven by naval demands, unlike in Britain, where it was mainly to meet the commercial challenges of merchant shipping.[61]

The US shipbuilding industry gained momentum towards the end of the nineteenth century, adopting scientific methods from Europe and amalgamating them with their chain production and outsourcing models. Since the US had a limited role in World War I, the increased demand for shipbuilding was largely in the merchant ship segment, and this was addressed by the creation of a federal agency that took control of all the shipyards and directed increased production. However, there was surplus production, leading to a slump after the war.

During World War II, there was greater demand for enhancing the production of merchant ships, since large numbers were being sunk by the Germans. The federal government set up the United States Maritime Commission (USMC) to build merchant ships and military support vessels. The navy was to coordinate the production of warships. Many actions were taken to meet the high production targets including the expansion of existing shipyards; the setting up of emergency shipyards; the standardisation of ships like *Liberty*-class and *Victory*-class; effective supply chain management; and innovative production technologies like the ones in the Kaiser Yards.[62]

The manufacturing of all equipment and machinery required for the ships was distributed to multiple vendors. The main engines, rudders, pumps, and motors were standardised and manufactured at different factories in assembly lines and dispatched to shipyards, as per the requirements, using an effective supply chain management system. Kaiser emergency yards[63] introduced innovative production technologies, borrowed from the automobile industry, to bring in transformation. Innovative techniques like Group Technology, pre-outfitting of sub-assemblies on the shop floor, welding in the down-hand position, introduction of production line concept, and organising the workplace to optimise production efficiency were put in place to achieve the fastest rate of construction. Using these innovations, the average time to build a 10,500 DWT Liberty class ship was reduced from 150 days to 28 days per ship.[64]

The US ascended to be the leader in global shipbuilding production and held that position from 1941 to 1945. At the end of the war, there was, however, a slump in demand and excess inventory necessitated the sale of surplus *Liberty-* and *Victory-*class ships to other nations at relatively low prices.[65] In subsequent years, the US share in shipbuilding decreased substantially, and the industry focussed primarily on retaining warship-building technology, with lead in the military-industrial complex.

The Post-World War II Period

After World War II, there has been a noticeable change in the global shipbuilding trends that deviate from the established pattern of the preceding centuries. Whilst the centre of world shipbuilding has shifted to some of the countries in maritime Asia, leading shipbuilding nations are no longer the leading sea powers of the world. Larger geopolitical developments and the interplay of economics and strategy facilitated the shift of shipbuilding base from the West to destinations in Asia. The theoretical context of technological power shift in shipbuilding is examined next.

NOTES

1   The carrying capacity of a ship equals the weight of water it displaces, and since water has a high intrinsic weight (64 pounds per square foot), ships carry large volumes of cargo.
2   JG Crowther and R Whiddington, *Science at War*, (His Majesty's Stationary Office, 1947), 152.
3   KN Chaudhuri, *Trade and Civilisation in the Indian Ocean: An Economic History from the Rise of Islam to 1750*, (Cambridge: Cambridge University Press, 1985), 15.
4   WJ Nichols and Celine Cousteau, *Blue Mind: The Surprising Science That Shows How Being Near, In, On, or Under Water Can Make You Happier, Healthier, More Connected, and Better at What You Do*, (New York: Little, Brown and Company, 2014), 9.
5   "Designing and Building a Wooden Ship," http://penobscotmarinemuseum.org/pbho-1/ships-shipbuilding/designing-and-building-wooden-ship, accessed 18 January 2019.

6   Todd and Lindberg, *Navies and Shipbuilding Industries*, 10–11.

7   Randhawa, *Indian Shipbuilding*, ch.1, n.3.

8   Usually defined as a measure of the capacity of a nation to use the sea in defiance of an adversary.

9   Chaudhuri, *Trade and Civilisation in the Indian Ocean*, ch.1, 59–61.

10   Kennedy, *The Rise and Fall of the Great Powers*, ch.1, 36.

11   Daniel Moran, "The international law of the sea in a Globalized world," *Globalization and Maritime Power*, ed. SJ Tangredi (Washington D.C.: NDU Press publications, 2002), 221-240.

12   Immanuel Maurice Wallerstein, *The Essential Wallerstein*, (New York: New Press, 2000), 87.

13   Kennedy, *The Rise and Fall of the Great Powers*, ch.1, 34.

14   Kevin L Falk, *Why Nations Put to Sea: Technology and the Changing Character of Sea Power in the Twenty-First Century* (New York: Garland Publishing, Inc., 2000), 1.

15   Alfred Thayer Mahan, *The Influence of Sea Power Upon History, 1660-1783*, (http://community.ebooklibrary.org/eBooks/WPLBN0000870131-The-Influence-of-Sea-Power-Upon-History-1660-1783-by-Mahan-A-T--Alfred-Thayer-.aspx?&Words=A%20T%20Mahan), accessed 15 June 2018, 1–2.

16   Mahan, *The Influence of Sea Power*, 26.

17   Chaudhuri, *Trade and Civilisation in the Indian Ocean*, ch.1, 12–14.

18   Chaudhuri, *Trade and Civilisation in the Indian* Ocean, ch.1, 15–16.

19   Chaudhuri, *Trade and Civilisation in the Indian Ocean*, ch.1, 14–16.

20   Chaudhuri, *Trade and Civilisation in the Indian Ocean*, 78.

21   Mahan, *The Influence of Sea Power*, ch.2, 27–28.

22   Mahan, *The Influence of Sea Power*, 327.

23   Alfred Thayer Mahan, *Influence of Sea Power Upon the French Revolution and Empire, 1793-1812*, (http://community.ebooklibrary.org/eBooks/WPLBN0000873225-Influence-of-Sea-Power-Upon-the-French-Revolution-and-Empire-1793-1812--Vol-2-by-Mahan-A-T—Alfred-Thayer-.aspx?&Words=A%20T%20Mahan), accessed 15 June 2018, 205–6.

24   Mahan, *Influence of Sea Power Upon the French Revolution and Empire*, 409-411.

25   Wallerstein, *The Essential Wallerstein*, ch.1, 85-86.

26   A form of agriculture in which the landowner allows the tenant to use the land in return for a share of the crops produced (the 'bataidaar' system in India).

27   Angus Maddison, *The World Economy: A Millennial Perspective* (Paris: OECD Publishing, 2001), 5–6.

28   Hedley Bull and Adam Watson, eds., *The Expansion of International Society*, (Oxford: Oxford University Press, 1985), 15–17.

29   Joseph A. Schumpeter, *Capitalism, Socialism and Democracy* (London: Routledge, 2013), 285.

30   Robert U Ayres, "Technological Transformations and Long Waves. Part I," *Technological Forecasting and Social Change*, vol. 37, no. 1 (March 1990): 1–37.

31   Mark Dunkley, *Ships and Boats: 1840 - 1950*, ed. Paul Stamper (Historic England, 2016), https://content.historicengland.org.uk/images-books/publications/iha-ships-boats-1840-1950/heag133-ships-and-boats-1840-1950-iha.pdf, accessed 10 September 2017, 1-2.

32   "Famous Ancient Chinese Ships, The Castle Ship Shipbuilding Techniques," *Ancient China Facts*, http://www.ancientchinalife.com/famous-ancient-chinese-ships.html, accessed 20 July 2018.

33   Sanjeev Sanyal, *Land of Seven Rivers - A Brief History of India's Geography* (Gurgaon: Penguin Books India Pvt Ltd, 2013), 120–21.

34   W. H. Moreland, "The Ships of the Arabian Sea about A.D. 1500," *Journal of the Royal Asiatic Society of Great Britain and Ireland*, no. 1 (January 1939): 63–74.

35   Jeffrey Hays, "Minoans (3000 B.C. to 1400 B.C.): Their Art, Culture and Religion and the Thera eruption," *Facts and Details*, http://factsanddetails.com/world/cat56/sub366/item2043.html, accessed 13 July 2018.

36   Anne Marie Smith, "Phoenician Ships: Types, Trends, Trade and Treacherous Trade Routes," *Dissertation, University of South Africa*, 2012, http://uir.unisa.ac.za/handle/10500/10344, accessed 10 September 2017.

37   "Byzantine Empire - Ancient History," *History.com*, http://www.history.com/topics/ancient-history/byzantine-empire, accessed 20 July 2017.

38   Deno John Geanakoplos, *Constantinople and the West: Essays on the Late Byzantine (Palaeologan) and Italian Renaissances and the Byzantine and Roman Churches* (Wisconsin: Univ of Wisconsin Press, 1989), 3–5.

39   George Saliba, *Islamic Science and the Making of the European Renaissance* (Massachusetts: MIT Press, 2007), 22–24.

40   Maddison, *The World Economy*, ch. 2, 57.

41   Roger Crowley, "Arsenal of Venice: World's First Weapons Factory," *Military History* vol. 27, no. 6 (March 2011), 62-70.

42   Maddison, *The World Economy*, ch.2, 54-58.

43   Maddison, *The World Economy*, 23.

44   Maddison, *The World Economy*, 65.

45   Chaudhuri, *Trade and Civilisation in the Indian Ocean*, ch.2, 139.

46   Colin Turner and Debra Johnson, *Global Infrastructure Networks: The Trans-National Strategy and Policy Interface* (Cheltenham: Edward Elgar Publishing, 2017), 16-17.

47   Chaudhuri, *Trade and Civilisation in the Indian Ocean*, ch.2, 140.

48   Small highly manoeuvrable sailing ships developed by the Portuguese to explore the west coast of Africa.

49   Maddison, *The World Economy*, ch.2, 59–66.

50   Maddison, *The World Economy*, 77–90.

51   WH Moreland, "The Ships of the Arabian Sea about A.D. 1500," *Journal of the Royal Asiatic Society of Great Britain and Ireland*, no. 2 (1939), 173–92.

52   Violet Barbour, "Dutch and English Merchant Shipping in the Seventeenth Century," *The Economic History Review*, vol. 2, no. 2 (January 1930), 261.

53   Kennedy, *The Rise and Fall of the Great Powers*, ch.1, 250–53.

54   David Lyon, *The Ship: Steam, Steel and Torpedoes: The Warship in the 19th Century* (London: Her Majesty's Stationary Office, 1980).

55   William H Thiesen, *Industrializing American Shipbuilding: The Transformation of Ship Design and Construction, 1820-1920*, (Gainesville: University Press of Florida, 2006), 16.

56   Robert K. Massie, *Dreadnought: Britain, Germany, and the Coming of the Great War* (New York: Ballantine Books, 1992), 468–97.

57 Antony Preston, *The Ship. Dreadnought to Nuclear Submarine* (London: Her Majesty's Stationary Office, 1980).

58 Todd and Lindberg, *Navies and Shipbuilding Industries*, ch.1, 101–103.

59 Todd and Lindberg, *Navies and Shipbuilding Industries*, 104.

60 Slaven Anthony, *The Shipbuilding Industry: A Guide to Historical Records*, ed. L. A. Ritchie (Manchester: Manchester University Press, 1992), 9–13.

61 Thiesen, *Industrializing American Shipbuilding,* ch.1, 167.

62 Frederic Chapin Lane et al., *Ships for Victory: A History of Shipbuilding under the U.S. Maritime Commission in World War II*, (London: Johns Hopkins University Press, 2001), 5.

63 Seven major shipbuilding yards owned by Kaiser Shipbuilding Company were located on the west coast of the United States during World War II. These shipyards were created in 1939 by American industrialist Henry J Kaiser.

64 S Navaneetha Krishnan, *Prosperous Nation Building Through Shipbuilding*, (New Delhi: KW Publishers Pvt Ltd, 2013), 71–72.

65 Tim Colton and La Var Huntzinger, "A Brief History of World Shipbuilding in Recent Times" (Alexandria: The CNA Corporation, 2002), http://www.dtic.mil/dtic/tr/fulltext/u2/a409101.pdf, accessed 29 July 2018, 6.

3

Context of Technology Power Shifts in Shipbuilding

Introduction

The Asian Maritime Crescent is a hub of erstwhile civilisational states with diverse ancient maritime legacies. These incredible legacies have been enshrined in their respective histories, dating back to the far-sighted maritime vision of the Song (960 to 1279) and Ming (1368 to 1644) dynasties of China, the Chola kings in India (330 BCE to 1279 AD), the great Japanese chronicles Kojiki and Nihon Shoki (cover the period between Japan's mythical origins to 697 AD), to the Goryeo and Joseon dynasties of Korea (918 to 1897). The strategic cultures of China, India, Japan, and Korea draw their political theories of statecraft from the respective ancient schools of thought – epitomised in the eras of their antiquity and golden ages that had firm maritime foundations – including Shih (Sun Tzu), Mandala (Kautilya), Bushido (Yamaga Takasuke),[1] and Yi Sun-Sin's innovative design- and use-of the *Geobukseon*[2] in sixteenth century Korea.[3]

However, as many Asian powers did not participate in the Industrial Revolution, their relative share in global manufacturing declined significantly. In 1750, the relative share of the global manufacturing of China and India was 32.8 and 24.5 per cent respectively, and by 1900, it had reduced to 6.2 and 1.7 per cent respectively. During the same period, Europe's global manufacturing increased from 23.2 per cent to 62 per cent.[4] Towards 1750, or the beginning of the Industrial Revolution, the global population grew exponentially. The population of Europe grew from 140 million to 266 million by 1850. During the same period, the population of Asia increased from 400 million to 700 million.[5] The rate of increase in the global population continued to be remarkably high and is projected to be 8.5 billion by 2030.[6] Asia and Africa are adding more numbers compared to Europe in this regard.

In the latter half of the seventeenth century, a 'Financial Revolution' started in Europe, bringing about a structured system of banking and credit.[7] Major European powers utilised the system of credit to pay for their wars.[8] This also enhanced ease of business and increased productivity, improving the standard of living. However, since the population of Asia was considerably higher, the share in the global manufacturing output was still dominated by Asian countries – such as India and China – though productivity levels were beginning to rise in sparsely populated Europe. In the early decades of the nineteenth century, the impact of the Industrial Revolution became more pronounced in Europe, increasing per-head production manifold. Asian countries that missed the Industrial Revolution began to lag behind Europe. As European powers set up colonies in Asia, they de-industrialised their colonies by abrogating trades, like textiles, to ensure the European manufacturers' complete dominance of the market.

In a turn of events, during the second half of the twentieth century, Asia was the fastest-growing economy, outperforming all other regions. This is in stark contrast to the previous four-and-a-half centuries when Asia had stagnated whilst other regions progressed. In 1500, Asia accounted for 65 per cent of the world's GDP, which reduced to 18.5 per cent in 1950. Since 1950, the Asian share has doubled.[9] This increase in Asia's relative contribution to the world's products – compared to the West – can be viewed as the 'recovery of Asia', or a power transition from the West to the East.[10]

Power Transitions

In the early seventeenth century, Western Europe emerged as the centre of power – or the 'core' of the modern world system. The Great Voyages of the late fifteenth and early sixteenth century ushered in a capitalist-based modern world system, with an increase in global trade. Trade volumes expanded to unprecedented levels. The Financial Revolution – in the latter half of the seventeenth century – facilitated business through a structured system of banking, credits, and exchange. The traded commodity diversified and for the first time, included 'staples' such as oil and grain in bulk, as well as coal and iron ore. To boost agricultural growth, leading powers engaged in the slave trade. After the Industrial Revolution, volumes of trade increased further; leading industrial nations felt the need to expand markets, dominate technologies, and have a captive resource base for raw materials for their industries. European powers found colonisation to be the best solution to meet the requirements of captive sources of raw materials and the market for finished products. Whilst the slave trade had a major impact on Africa, Asia was colonised. However, both developments remained intricately linked to the power transitions in Europe.

The theory of power transition examines shifts of power and causes of conflicts among nation-States.[11] An industrialising nation enhances its comprehensive national power and its ability to influence other nations in the international system, thereby going through a power transition. This typically happens in three stages: a nation begins to industrialise and becomes a potential power; it then goes through a period of transitional growth; and finally, attains a state of power maturity. Power transition, therefore, is intrinsic to the industrial growth of a nation. An emerging power makes efforts to realign the existing power equation of the world order to stake claim to its legitimate position. Whenever a leading power of the existing international system enjoys a preponderant position, the status quo is maintained, and peaceful adjustments are made. However, when the emerging challenger gains comparable power, there is a great power conflict and this leads to the readjustment of the international order and the establishment of a new hierarchical system.[12] Two power transitions in the preceding centuries have had a significant impact on the evolution of 'Maritime' Asia – the first was the emergence of Britain as a 'hegemon' after the Industrial Revolution, while the second was the emergence of the US as a superpower after the Second World War.

The Rise of Great Britain

In the seventeenth century, the Dutch Republic was the dominant power in Europe. It enjoyed clear maritime and commercial supremacy and monopoly of trade with Africa, Asia, and the Americas. This was challenged by the rising powers, England and France, which were contending for dominance in the eighteenth century.[13] There were seven major Anglo-French Wars between

1689 and 1815 to consolidate supremacy in Europe. Both England and France faced significant problems in generating finances for the wars, yet sustained high economic growth that was required to ascend to the lead position. Britain succeeded in implementing a financial system, which facilitated its lead over France.

To support the Anglo-French Wars, English war expenditure requirements increased tremendously from 1688 to 1815. Hence, there was a shortfall of the order of 33 per cent between the public income and total expenditure.[14] The government had the option of increasing taxes from the citizens, inflating the currency, or writing off accumulated debts. All these options would have proven to be self-defeating, since taxes were already extremely high, and the economy was not very robust. Therefore, the government formulated an organised system of long-term government borrowing. This proved to be the best way to tap the wealth of a nation which was at a crucial stage of development, with minimum additional burden on its citizens. The French failed to match up on the system of generating funds, and this had a disastrous effect, leading to their ultimate defeat.

Britain's victory in the initial Anglo-French Wars of 1739 to 1763 ensured that it could preserve and extend its markets in North America and India and increase its exports. If these markets had fallen to the French, English economic development would have retarded. An equally important effect of the rise of public borrowing in the first half of the eighteenth century was that it created a range of securities in which mercantile and financial houses could invest or disinvest easily. London's new banks, insurance offices, trading companies, merchants, brokers, and workers found varied and flexible investment opportunities than the earlier option of investments in land alone. This facilitated the creation of a complex structure of services which could not have been created otherwise since there was no industrial sector whose bonds could have been used for the same purpose. Any delay in the emergence of London as a financial centre would have resulted in a failure to hold existing overseas markets and win new ones, thereby setting back England's industrialisation and changing the course of European history.[15]

The Financial Revolution[16] established Britain in Europe's lead position; it was a crucial factor that led to Britain and its allies' victory over the French in the seven major Anglo-French Wars from 1689 to 1815. The government financed wars by borrowing through a system of selling bonds or long-term stock, paying interest to those who invested in these bonds or stocks. The capital that was available through this form of long-term borrowing was then used by the government to make payments to military contractors, shipbuilders, provision merchants, and the armed services.[17]

The two-way system – raising funds while simultaneously spending vast sums of money – became the blueprint for the future growth story. This process boosted development and led to the Industrial Revolution. Major innovations in technology, specifically in steam engines and textiles in the 1760s, further consolidated Britain's position as a leading power. In 1783, the defeat of the British in the American War of Independence led to thirteen colonies of the British Empire declaring independence. This pushed the British to focus on firming up its control over colonies across the world. Finally, the Napoleonic Wars (between 1792 and 1815) – when British interests triumphed over the French — positioned the British Empire as the 'hegemon'. As the dominant sea power, the British expanded their area of influence to Asia and beyond.[18]

In the latter half of the nineteenth century, Britain translated its lead in the Industrial Revolution by introducing several technological improvements in military applications. There were improvements in the muzzle-loading gun, which included percussion caps, rifling etc.; the breech-loader vastly increased the rate of fire, leading to the introduction of the Gatling or rapid-fire guns like the Maxims; and light field artillery ushered a new 'fire power' revolution. Along with the revolution in guns, they introduced steam-driven gunboats with high mobility and major ironclad warships with high firepower. These developments by the British cemented their lead and dominance at sea.[19] The integration of economics, technology, and military capability became a standard characteristic of the new industrial powers that emerged in the latter half of the nineteenth century.

The Rise of New Industrial Powers

By the latter half of the nineteenth century, the global communication network was well-established, facilitating trade and quicker information exchange across the world. Advancements in fields like telegraphs, steamships, railways, and the printing press provided seamless connectivity. As a result of newfound global connectivity, innovations in technology and advancements in manufacturing and production in the industrially-advanced countries began getting transferred across continents very quickly.[20] In 1879, for instance, Gilchrist and Thomas invented a way to turn cheap phosphoric ores into basic steel. Within five years, there were 84 basic converters in operation in western and central Europe[21] and the process had reached other continents as well. During this period, the introduction of electricity, steel, and mass production also transformed shipbuilding technologies.

Japan, Italy, and Germany emerged as new industrial powers. Japan emerged from its self-imposed isolation after the 'Meiji Restoration' in 1868. Italy and Germany were unified in the 1870s. Since this was a period during which there was increased application of technology in the military, these industrial powers also emerged as military powers. The steel output of these countries increased substantially, facilitating a significant increase in their military potential. By the end of the nineteenth century, each of these powers was acquiring territory overseas, building a modern navy to complement the standing armies, and forming alliances at strategic levels with one of the established world powers.

There was simultaneously an industrial and agricultural expansion in the United States and a military expansion in Russia. A measure of the industrial potential of the leading countries and their relative share in world manufacturing during this period provides a fair assessment of the relative status of these countries in the global hierarchy. Accordingly, Tables 3.1 and 3.2 bring out the shift in the positions of the powers between 1880 and 1938, considering the total industrial potential in relative terms and relative shares of world manufacturing.

Table 3.1: Total Industrial Potential of the Powers in Relative Perspective, 1880-1938 (UK in 1900 = 100)[22]

	1880	*1900*	*1913*	*1928*	*1938*
Britain	73.3	[100]	127.2	135	181
US	46.9	127.8	298.1	533	528
Germany	27.4	71.2	137.7	158	214
France	25.1	36.8	57.3	82	74
Russia	24.5	47.5	76.6	72	152
Italy	8.1	13.6	22.5	37	46
Japan	7.6	13	25.1	45	88

Table 3.2: Relative Shares of Global Manufacturing Output,
1880-1938 (per cent)[23]

	1880	1900	1913	1928	1938
Britain	22.9	18.5	13.6	9.9	10.7
US	14.7	23.6	32.0	39.3	31.4
Germany	8.5	13.2	14.8	11.6	12.7
France	7.8	6.8	6.1	6.0	4.4
Russia	7.6	8.8	8.2	5.3	9.0
Italy	2.5	2.4	2.4	2.7	2.8

A New World Order

The new industrial powers also acquired substantial military capability, which became a significant element in the diplomatic calculus of the time. Each of these powers became alliance partners to an older power. The international struggles now were much more intense compared to the limited and localised power clashes of nineteenth-century Europe. However, the most notable change was in the relative share of global manufacturing as depicted in Table 3.2 above. The coupling between technology and industrial base on one hand, and military power and diplomatic influence on the other, became even more significant. Diplomacy began to play a significant role in shaping the sphere of technology-economics-strategy. A spin-off of this was the emergence of the 'technology denial' regimes – an effective technology strategy and an effective instrument of statecraft. The decisive factor during a conflict was the ability of the competing great powers to reorganise the productive forces to respond with an unprecedented surge in industrial production to support the war-fighting capabilities.[24]

The multipolar world of 1885 was replaced by a bipolar world as early as 1943. The post-World War II bipolar world and the recalibration of great power equations had a significant impact on Asia. As the age of empires and colonies came to an end, many independent states were formed – open to influences from either of the established power centres. These new independent entities including India, China, Indonesia, Philippines, Iran, Taiwan, the Baltic states, Finland, Hungary, and Romania,[25] had to choose their type of economy, form of governance, and above all, the power they would lean towards. Each of them was to make the choice while pursuing their comprehensive national interest, with the state of their economy significantly influencing their choices. The bottom line, however, was that many of the developments that followed were hugely impacted by the balance of power between the two superpowers.

Balance of Power

As the Second World War was ending, each of the major victorious powers – Great Britain, the US, and the USSR – proposed a plan of action that would ensure enduring world peace. Each of these proposals was profoundly influenced by their own historical experiences. Winston Churchill, for instance, wanted to reconstruct the traditional balance of power in Europe by rebuilding Great Britain, France, and Germany, which – along with the US – could counterbalance the Soviet Union. Stalin's approach was rooted in the traditional Soviet foreign policy of expanding Russian influence in Central Europe, turning the countries conquered by the Soviet armies into buffer zones as protection against future aggressions. Roosevelt envisaged the post-war order as one consisting of a board of directors of the "Four Policemen" – the US, the USSR, Great Britain, and China – enforcing peace against any defaulting or miscreant power.[26]

The ensuing events from early 1945 to the end of the war – as well as the succession of Churchill by Atlee in Britain and Roosevelt by Truman in the US – influenced the 'great power' engagements. More specifically, the views of the US and the USSR were most significant to the emerging world order – largely bipolar. During the successive rounds of summits, the Western powers continued to insist on fair and transparent elections and the establishment of democratically-elected governments in all the States, including Finland, Hungary, Romania, and Bulgaria. The Soviet Union was suspicious of the West's intentions; it stuck to its strategy of firm control over border states. The Western powers concluded that Soviet influence was expanding, considering the events in Greece and Turkey,[27,28] as also the developments in Cuba, Yugoslavia, and Poland.[29,30]

The growth and expansion of the Soviet Union appeared to confirm the geopolitical predictions of Mackinder and others – a gigantic military power would control the resources of the Eurasian heartland; further expansion of that State into the periphery – or 'Rimland' – would need to be contested by the 'great maritime' States if they were to preserve a global balance of power.[31]

In 1946, based on approach papers by H Freeman Matthews and others, the US concluded that it enjoyed military superiority at sea and air, whilst the Soviet Union had significant supremacy on land. Considering their weakness on land within the Eurasian landmass, the Americans planned to restrict the use of force to those areas where the Soviet armies could be countered defensively by naval, amphibious, and air power. This strategic intent was to define the continued significance of sea power in the future great power rivalry. The Americans also decided that use-of-force was best executed under the charter of the United Nations rather than through a unilateral action.[32]

George F Kennan and others formulated the US policy of containment,[33] limiting its scope to protecting major industrial power centres from Soviet expansion. These centres were identified as Western Europe and Japan. Subsequently, the scope of the US policy of containment was expanded to all free institutions of the Western world.[34] The core of American thinking was rooted in two lessons from its recent history. First, a belief that threats to political stability arise primarily from gaps between economic and social expectations and reality, hence the Marshall Plan. Second, the best protection against aggression is to possess overwhelming power and a political will to use it, leading to the establishment of the Atlantic Alliance.

The Marshall Plan was designed to get Europe on its feet economically and the North Atlantic Treaty Organization (NATO) was to look after its security.[35] The Marshall Plan included massive economic assistance and the transfer of technologies to build infrastructure and industry in specific regions, like Western Europe and Japan. Shipbuilding formed a major component of the transfer of technology to Japan. In the following years, both major powers consolidated their allies. In Greece and Turkey, where the British could no longer pursue influence, the US took up the mantle and similarly got engaged in many other regions to counter the spread of communism. Initially, the Marshall Plan did not include the Korean Peninsula, however, the attack by North Korea on the South was seen as the communists' expansion, beyond China, which needed to be contained. Consequently, the US entered the Korean War with decisive might and subsequently got involved in Vietnam.[36]

In the ensuing period, there was a constant counterbalancing between the Soviet Union and the US, which had established themselves as the world's two leading power centres. However, there were some major differences between these powers. Whilst the Soviet Union had borne the brunt of World War II, the US largely remained a distant participant with its own soil intact. In the Soviet Union, industrial and technological infrastructure had been grossly damaged and required massive reconstruction effort, whereas, in the US' case, the productive capacity was fully functional. The US enjoyed a clear advantage over other powers in the world, in terms of capacity to finance growth. It had an open and transparent system that was fuelled by independent research and scholarship, partnered with private entrepreneurship. This made the US an attractive destination for talents from the academic world to further accelerate growth. The Soviets, by contrast, had a controlled economy and a closed system with much lower standards of living.[37]

The natural choice for the two major powers was therefore quite clear. The Soviets worked on expanding their realm of influence by military dominance and aiding popular movements against governments, based on economic shortfalls in the expectation of the public in various countries. The US, on the other hand, extended loans and grants for infrastructure build-up and industrial development, and opened markets to the allies to facilitate growth. The transfer of shipbuilding technologies, especially to Japan, was significant. Since the opening of markets provided growth stimulus to global trade – largely sea-borne, there was an increase in demand for merchant shipbuilding to expand. The US also continued to invest in defence for dominance in the military domain, especially its naval power. In Asia, Japan was central to the US' plans for rebuilding efforts immediately after the war.

The Rise of Japan: Shipbuilding as the Enabler

After World War II, the US and its allies commenced the reconstruction of Japan. Initially, the aim was to transform and demilitarise Japanese society. Subsequently, the aim was modified to rebuild the economy through giant firms called '*zaibatsu*', like Mitsubishi and Sumitomo. The creation of Japan's own military was encouraged to ease the economic and military burden of the US, while ensuring that Japan becomes an anti-communist bastion in Asia.[38] The reforms resulted in an economy that was run as a corporate with government support, gradually leading to wealth being transferred from the aristocracy to the middle class.[39] The government implemented land reforms, took over land holdings from aristocrats, abolished titles of the aristocracy and large landowners, and sold the land to tenants at reasonable rates. These reforms ended feudalism and led to land ownership by citizens.

Some other global developments after World War II also helped in the rapid reconstruction and economic growth of Japan. Since many countries were decolonised, Japan gained access to a larger market. Until then, the colonies had been restricted to trade with just their erstwhile masters. The world economy shifted to the Bretton Woods system[40] of monetary management, which reduced tariff barriers and facilitated free trade as the foundation of the international economy.

Under the Marshall Plan, the US provided financial assistance to Japan to rebuild and develop modern industrial infrastructure and manufacturing. Systemic reforms discouraged imports and facilitated a low-cost, export-oriented economy. Whilst financial aid and reforms facilitated reconstruction, the will of the Japanese to create 'made in Japan' products, and their work ethos and culture contributed immensely to the success of what came to be called the 'Japanese miracle' or the 'Japan Inc.'[41] The Japanese implemented innovations in the workplace, organised production floors and supply chains, and soon emerged as world leaders. Subsequently, the Korean War offered a great opportunity for the Japanese economy to gain from the transit and sourcing for US military requirements.

Viewed over the decades, the Japanese economic rise happened in three stages. The first stage focussed on coal- and steel-production and heavy industries, like shipbuilding and timber, in the 1950s. In the second stage – during the 1960s and 1970s, the focus was on consumer products and automobiles for export. The third stage was the development of knowledge-based products like computers and electronics. The success of this model made it the *de facto* blueprint for all the emerging powers – especially those in maritime Asia – to follow, with suitable adjustments for country-specific and time-specific factors.[42]

Shipbuilding was a major industry whose know-how was transferred from the US to Japan in the first stage of reconstruction. Japan had an advanced shipbuilding industry in the pre-war era and most of its shipbuilding infrastructure had escaped damage during the war, and was therefore largely intact and usable. The US shipbuilding industry, which had built around 5,777 cargo ships during World War II, was struggling with excess capacity and unsold inventory of around 1000 ships, including commercial vessels and warships. The US government had to close 31 shipyards because of a fall in demand and excess capacity. The shipbuilding industry, therefore, was amenable to transition to regions where a skilled workforce was available at a lower cost. The US government decided to move away from the shipbuilding industry, but retained warship building under state control.

The US identified shipbuilding as a preferred sector for the transfer of technology to Japan, considering the stage at which shipbuilding was in the industry life cycle. The industry life cycle theory defines distinct stages of any industry: inception, growth, maturity, and decline. The stages of inception and growth are technologically intensive and involve innovations in the process formulation and experimenting with production chains. These phases demand a technically skilled workforce and require capital investment. The industry is considered to have matured when the business processes and the production methods have been standardised, mass or bulk production has stabilised, and profits are maximised – meeting the demand.[43] At this stage, the industry is most amenable to moving to regions with abundant low-cost skilled workforce. The shipbuilding industry was at a mature state in the US after the Second World War.

The US had expanded its capacity to build ships to meet surges in demand during World War II. Capacity expansion was achieved using 'emergency shipyards', which were built in a short time span, implementing novel and disruptive changes to ship design and construction. This was largely credited to pre-war planning – often referred to as the 'Long Range Program' by the United States Maritime Commission (USMC), as well as the transformative changes in production methods by the Kaiser emergency shipbuilding yards. These shipyards were owned by Henry J Kaiser.[44]

Kaiser had no prior experience in shipbuilding. He used his lack of shipbuilding experience to try out new methods in production lines, based on the success of similar methods in the automobile industry. He pioneered the introduction of supply chain management to streamline the timely availability of equipment and machinery, the concept of Group Technology (GT), welding in the down-hand position, as well as the layout of shipyards based on the production line concept. The governing principle was to organise the work to suit the worker. These shipyards could construct ships in the shortest possible time. For example, the average time to build a *Liberty*-class ship was reduced from 150 days to 28 days. Once the US decided to assist in rebuilding Japan, the government facilitated the movement of many of the workers who worked in the Kaiser yards to Japan, laying the foundation for the Japanese shipyards.[45]

The shipbuilding recovery plan was formulated in Japan and funded by the US and Japan Development Bank. National Bulk Carriers (NBC) – an American shipping company owned by D K Ludwig – took Kure shipyard in Japan on lease and implemented industrial engineering principles that had been developed by the US during the war. The techniques used in the Kaiser shipyards were also implemented. Dr W Edwards Deming – a renowned American statistician – who went to Japan after World War II, introduced the Statistical Control Methods (SCM) in Japanese shipbuilding. Dr Hisashi Shinto, also known as a founding father of Japanese shipbuilding, introduced the flow-line method – a combination of block construction and project-scheduling techniques, helping Japanese shipbuilding to take on the leading position.[46]

The 1956 Suez Crisis forced oil transportation from the Persian Gulf to the European nations to go around Africa, which required additional time and hiked costs. To transport larger quantities in each trip and achieve economies of scale, there was a huge demand for large tankers. During this time, European shipyards had the infrastructure to build ships smaller than 32,000 DWT; only Japanese shipyards were building large tankers since they were being used to transport oil to Japan – in the absence of any canals in the transit route. The Japanese shipyards responded promptly to the market demand and made the best use of the available skilled manpower, capturing the maximum global shipbuilding market share.[47] Japan became the world's leading shipbuilder by the 1960s.

The Rise of Maritime Asia: Enabled by Shipbuilding

Between 1965 and 1995, South Korea, Taiwan, Hong Kong, and Singapore – often referred to as the 'Four Tigers' – raised per capita incomes six-fold. Indonesia, Malaysia, and Thailand tripled their income levels during the same period.[48] The export-led industrialisation of South Korea was facilitated by substantial aid from the US after the Korean War, and subsequently from Japan in the 1960s. It focussed on an export-driven economy with heavy investments in technology-intensive industries such as shipbuilding, steel, petrochemicals, and electronics. The Vietnam War came as a great economic opportunity, through the supply of the aforementioned war-waging material wherewithal for the US' military. Park Chung-hee emerged as the leader in South Korea in 1961, normalising ties with Japan and paving the way for Japanese investments. These investments, together with local savings and American aid, ensured that the economy firmly pursued reforms and achieved levels of growth unprecedented in history.[49]

The 1973 OPEC crisis resulted in reduced crude production and led to a drastic fall in the global demand for oil tankers. This triggered a downturn in the Japanese shipbuilding industry, forcing reduced production and massive restructuring. The rise of shipbuilding in South Korea during this period and the availability of skilled manpower, lower currency value, as well as the cost of steel favoured competition to tilt towards it. This was also facilitated by the US' commitment to South Korea after the end of the Korean War in 1953. By the 2000s, South Korea emerged as the world leader in shipbuilding.[50]

Next to follow was the rise of China in Asia. The formation of the People's Republic of China in 1949 and the establishment of a communist state was perceived as a major loss for the US since it went into the Soviet fold.[51] Differences between China and the Soviets grew over the years and led to their breakup in the late 1950s. President Nixon travelled to China in 1972 to normalise relations and open the channels for American investments to aid Chinese growth.[52] However, until then and till 1978, largely under Mao Zedong, China averaged 3 per cent growth per year. During this time, China experienced declining productivity, recurring food crises, and little improvement in the living standards of its people. The Great Leap Forward[53] failed and the Great Famine of 1959–1961 affected its growth story. After 1978, under the leadership of Deng Xiaoping, the strategy of growth-based capital accumulation took off through '*Gaige Kaifang*' or 'reform and opening up'. This was gradual, incremental, and de-centralised. Between 1998 and 2007, there was a fillip to privatisation and trade liberalisation. China sustained average yearly growth of 8 per cent for the next three decades, emerging as the second-largest economy in the world while being home to nearly 20 per cent of the global population.[54]

The Chinese entered the shipbuilding market in the 1970s, during Deng Xiaoping's regime. The Chinese shipyards initially obtained technological expertise from Japan and South Korea.[55] Availability of cheap labour, structural reforms in finance and management, government policies to support production and export, and management of demand contributed to the growth of shipbuilding as a strategic industry, leading to China attaining global leadership in 2009. The Chinese initially focussed on low-end technologies to increase market share; once they were established as a major player, they successfully worked out consultancies with international majors in high-end shipbuilding like the LNG carriers and major international ancillary manufacturers. From 2002 to 2012, commercial shipbuilding output surged thirteen-fold. China achieved its set goal of becoming the world's largest shipbuilder by 2015.[56]

India implemented important reforms in the 1990s by opening its economy. From 1991 to 2003, India's GDP grew at an average of 5.5 per cent a year. This was a growth acceleration of 1 percentage point a year over the previous two decades, and was followed by a high growth from 2004 to 2008 when the average annual growth rate was 8.8 per cent. Although there was a slowdown following the global financial crisis in 2008–2009, the economy slowly reverted to improved growth rates and stabilised. India's growth has been well diversified; the acceleration of value added has been the fastest in services, followed by industry. Within the service sector, a large part is accounted for by modern services, comprising financial services, communications, and the IT sector. The overall growth has not just accelerated but has also become more stable over time.[57]

The Emergence of Maritime Asia as a Manufacturing Hub

As some of the maritime Asian countries became leading shipbuilders of the world and their combined capacity virtually captured the entire global demand, they also became major economies of the world. As expected, the growth was technology-driven with a substantial contribution from the shipbuilding industry. This trend has been substantiated throughout history, and there has been good evidence of a strong correlation between economic growth and an increase in the share of global shipbuilding of maritime nations. The trend has been more pronounced in the latter half of the nineteenth century, as can be seen in Table 3.3.

Table 3.3: **Global shipbuilding leaders**[58]

Year		Country	Market share (Delivery in GT) %
From	*To*		
Early 1860	1940	United Kingdom	80
1941	1945	United States	90
1946	1960	Europe (UK, Germany and Italy)	65
1961	1999	Japan	50
2000	2009	Republic of Korea (ROK)	38
2010	Onwards	China	39

One of the primary reasons why shipbuilding causes economic rise is because it facilitates the growth of skilled manpower and a good technology base. For example, British shipbuilding peaked in 1920 and produced 1.9 million gross tonnages, employing nearly 326,000 workers in 135 major shipyards. Meanwhile, American shipbuilding production peaked in 1943 and between 1939 to 1945, roughly 50 million DWT of large cargo carriers and tankers were built and delivered by the US. These consumed around 25 million tons of steel and engaged nearly 640,000 workers at the peak of employment.[59] The growth of shipbuilding also leads to the setting up of manufacturing bases for engines, alternators, rudders, pumps, motors, sensors and systems like radars, navigation aids, etc. It creates demand for a vast spectrum of technologies and employs many workers, enabling broad-based economic growth.

Once the shipbuilding and ancillary industries as well as the steel, sheet metal, railroads, automobiles, electricity, aerospace, and consumer goods industries were established in the maritime Asian countries, this region emerged as the manufacturing hub of the world. The newfound seamless connectivity and ease of communication led to faster diffusion of new technologies from the West and the skilled workforce in these regions began to innovate and compete with the best products in the global market. Therefore, with advances in technology, the countries in maritime Asia consolidated their position as the global manufacturing hub.

During this phase of global technological transformation, a significant development was the dual use of civilian infrastructure for producing platforms and equipment for the military. In many of the maritime Asian countries, the modern shipyards that produced state-of-the-art merchant ships were used to build warships which were contemporary in design and capability. Some of the leading shipyards in maritime Asia began to build modern warships and submarines in collaboration with established weapons and sensor OEMs from the global military industrial complex. They emerged as world-class system integrators and warship builders who could compete globally.

The common thread in the overall growth story of the Asian countries – during the post-World War II period – was that each one of these initially benefitted from substantial financial assistance in the form of long-term loans and grants or investments; they all focussed on infrastructure creation and infusion of modern technologies; and each one embarked upon an export-oriented production market. In that sense, their growth has largely been technology-driven. The West, however, continued to innovate and introduce cutting-edge technologies like Information and Communication Technologies (ICT) that ushered in Industry 3.0 (Third Industrial Revolution). The know-how of these new technologies diffused to the emerging Asian region in a few years; by this time, the skill sets for absorbing these technologies were available.

The Rise of the Indo-Pacific

The economic transformation, industrial advancement, and military power of Japan, South Korea, China, and India after World War II brought to prominence the 'Indo-Pacific' as a region of intersecting interests among major maritime trading and strategic powers that shaped global events. The 'Indo-Pacific' coinage is the contemporary reference to the Asia-Pacific landscape,[60] bound strategically by India in the west, the US in the east, Japan in the north, and Australia in the south. This region encompasses over 51 per cent of the earth's surface and includes approximately half the world's population. The UN has predicted that by 2050, seven out of ten people will live in the Indo-Pacific region. It is, therefore, a natural consequence that the global order is being reshaped by the developments in this region. This region has emerged as the global centre of economic gravity, as a theatre of increasing security tensions, and as a source of decisive influence in global environmental change.

As shipbuilding contributed to the overall industrial and economic growth of the countries in maritime Asia, the collective share of the world's Gross Domestic Product (GDP) from the region increased substantially. Economic growth also made it possible for these countries to raise powerful militaries and a supporting military industrial complex. The magnitude and growth of the maritime Asian countries have led to the Indo-Pacific construct.

The rise of the countries in the Indo-Pacific region has been greatly influenced by technology, with manufacturing contributing significantly to economic gains. The onset of Industry 4.0 is shaping the politico-economic landscape, with the West employing all its efforts to continue to be the hub of innovation and centre for the next generation of technologies. The developments in these emerging technologies, as well as their impact on the shipbuilding industry, with a special focus on Japan, South Korea, China, and India, will be examined next.

NOTES

1   Missions and obligations of the Samurai.
2   A large warship used in Korea during the Joseon dynasty between the early 15th century and 19th century.
3   Vijay Sakhuja, *Asian Maritime Power in the 21st Century: Strategic Transactions: China, India and Southeast Asia* (Singapore: ISEAS, 2011), xi.
4   Kennedy, *The Rise and Fall of the Great Powers*, ch.1, 190.
5   Fernand Braudel, *Civilization and Capitalism, 15th-18th Century, Vol. I: The Structure of Everyday Life*, trans. Siân Reynold (Berkeley: University of California Press, 1992), 42.
6   "UN Projects World Population to Reach 8.5 billion by 2030, Driven by Growth in Developing Countries," *UN News*, 29 July 2015, https://news.un.org/en/story/2015/07/505352-un-projects-world-population-reach-85-billion-2030-driven-growth-developing, accessed 11 August 2018.
7   PGM Dickson, *The Financial Revolution in England: A Study in the Development of Public Credit*, 1688-1756 (New York: Routledge, 2017), 10–12.
8   Kennedy, *The Rise and Fall of the Great Powers,* ch.1, 98.
9   Maddison, *The World Economy - A Millennial Perspective,* ch.2, 142.
10 Joseph S Nye, Jr., "Understanding 21st Century Power Shifts," *The European Financial Review*, 24 June 2011, http://www.europeanfinancialreview.com/?p=2743, accessed 11 April 2018.
11 The theory postulates that the spread of industrialisation to different nations at different times and different rates provides a key to understanding fundamental patterns in international relations.
12 AFK Organski, "Power Transition," *International Encyclopedia of the Social Sciences*, http://www.encyclopedia.com/social-sciences/applied-and-social-sciences-magazines/power-transition, accessed 08 March 2018.
13 Maddison, *The World Economy - A Millennial Perspective,* ch.2, 96.
14 Kennedy, *The Rise and Fall of the Great Powers,* ch.1, 105.
15 Dickson, *The Financial Revolution in England,* ch.1, 11.

16   The main features of the Financial Revolution were the development of long- and short-term borrowing, the relations between the treasury and the city, the development of a market in securities, and the sources of capital invested in public loans.

17   Kennedy, *The Rise and Fall of the Great Powers*, ch.1, 100.

18   Kennedy, *The Rise and Fall of the Great Powers*, 180.

19   Kennedy, *The Rise and Fall of the Great Powers*, 192.

20   Kennedy, *The Rise and Fall of the Great Powers*, ch.1, 254.

21   David S Landes, *The Unbound Prometheus: Technological Change and Industrial Development in Western Europe from 1750 to the Present*, 2nd edition (Cambridge, UK: Cambridge University Press, 2003), 259.

22   P Bairoch, "International Industrialization Levels from 1750 to 1980", *Journal of European Economic History*, vol. 11, no. 2 (Fall 1982), 292-299.

23   Bairoch, "International Industrialization Levels from 1750 to 1980", 296-304.

24   Kennedy, *The Rise and Fall of the Great Powers*, ch.1, 250-253.

25   Kennedy, *The Rise and Fall of the Great Powers*, ch.1, 480-95.

26   Henry Kissinger, *Diplomacy*, Reprint edition (New York, NY: Simon & Schuster, 1995), 395.

27   In Greece, there was a communist rebellion guerrilla force which was threatening to overthrow the government. It was suspected that these forces were supported by the Soviet Union. In Turkey, the Soviets were demanding some form of control over Dardanelles, from where Turkey was able to dominate the strategic waterway from the Black Sea to the Mediterranean.

28   History.com Editors, "Truman Doctrine Is Announced - Mar 12, 1947," *History*, (13 November 2009), http://www.history.com/this-day-in-history/truman-doctrine-is-announced, accessed 11 August 2018.

29   Pro-communist governments were installed under the influence of the Soviet Union.

30   Kissinger, *Diplomacy*, ch.3, 443.

31   Kennedy, *The Rise and Fall of the Great Powers*, ch.1, 469.

32   Kissinger, *Diplomacy*, ch.3, 449.

33   Containment is a geopolitical strategy to stop the expansion of an enemy.

34   "Milestones: 1945–1952 - Office of the Historian," https://history.state.gov/milestones/1945-1952/kennan, accessed 23 March 2018.

35   Kissinger, *Diplomacy*, ch.3, 456.

36   Kennedy, *The Rise and Fall of the Great Powers*, ch.1, 469-70.

37   Kennedy, *The Rise and Fall of the Great Powers*, ch.1, 460-61.

38   Kennedy, *The Rise and Fall of the Great Powers*, ch.1, 492.

39   Jeffrey Hays, "Economic History after World War II in Japan, South Korea and South East Asia," http://factsanddetails.com/asian/cat62/sub408/item2560.html, accessed 12 August 2018.

40   "Creation of the Bretton Woods System," *Federal Reserve History*, https://www.federalreservehistory.org/essays/bretton_woods_created, accessed 01 April 2018.

41   Hays, "Economic History after World War II", ch.3.

42   Hays, "Economic History after World War II", ch.3.

43   "Industry Life Cycle," *Inc.com*, http://www.inc.com/encyclopedia/industry-life-cycle.html, accessed 31 July 2016.

44 Krishnan, *Prosperous Nation Building Through Shipbuilding*, ch.1, 71.

45 Krishnan, *Prosperous Nation Building Through Shipbuilding*, 74.

46 Krishnan, *Prosperous Nation Building Through Shipbuilding*, 86–87.

47 Krishnan, *Prosperous Nation Building Through Shipbuilding*, 89–90.

48 Hays, "Economic History after World War II", ch.3.

49 Hays, "Economic History after World War II", ch.3.

50 Krishnan, *Prosperous Nation Building Through Shipbuilding*, ch.1, 102-3.

51 Kissinger, *Diplomacy*, ch.3, 135.

52 Henry Kissinger, *On China* (India: Penguin Books India Pvt Ltd, 2012), 255.

53 A nationwide programme of economic collectivisation was launched in 1958. Mao Zedong outlined his vision of China in perpetual motion.

54 Xiaodong Zhu, "Understanding China's Growth: Past, Present, and Future," *Journal of Economic Perspectives* vol. 26, no. 4 (November 2012), 103.

55 Colton and Huntzinger, "A Brief History of World Shipbuilding in Recent Times", ch.1, 19.

56 Shannon Tiezzi, "Chinese Naval Shipbuilding: Measuring the Waves: An Interview with Andrew S. Erickson," *The Diplomat*, 25 April 2017, https://thediplomat.com/2017/04/chinese-naval-shipbuilding-measuring-the-waves/, accessed 16 April 2018.

57 World Bank, *India Development Update, March 2018: India's Growth Story* (World Bank, 2018), 13–16, https://doi.org/10.1596/29515, accessed 08 December 2018.

58 Krishnan, *Prosperous Nation Building Through Shipbuilding*, ch.1, 64.

59 Krishnan, *Prosperous Nation Building Through Shipbuilding*, 64-74.

60 Bagchi, "Peaceful Periphery on the Seas".

4

The Technology Conundrum: Transitions and Transformations

Introduction

During the post-World War II period, export-oriented economies grew in maritime Asia alongside major transformations in technology taking place in the world. Technology contribution is, therefore, central rather than peripheral in transforming the countries in maritime Asia. As Western advanced economies offshored manufacturing to Asian destinations and opened their markets for these countries, export-oriented economies in the Asia-Pacific began to take shape. This was a win-win situation for all; the companies in the West that invested in Asian manufacturing made huge profits, consumers got products at lower prices, and the Asian economies experienced exponential growth. Two consequential, though unintended, developments followed. First, the countries where manufacturing bases were set up began to see an increase in the pool of skilled workforce, which also began to innovate and invent more efficient production methods hitherto unknown to the West. Second, as the workforce in these countries swelled, there was an expansion of the middle class, which was the new local consumer base. This had a regenerative effect on the economy. The interplay between economics and technology evolved through successive Industrial Revolutions (IRs) that will be explained in the subsequent paragraphs.

Classification and Periods of Industrial Revolution (IR)

Although there is a broad consensus in referring to the invention of a group of technologies that was largely centred in Europe in the 1760s as the 'Industrial Revolution', many scholars have classified distinct periods when a group of technologies substantially changed the way of life, as Industry 1.0, Industry 2.0, Industry 3.0, or Industry 4.0 (1st, 2nd, 3rd or 4th Industrial Revolution). These classifications have been varied and diverse, depending on the criteria chosen for the study and hence there is no standard frame of reference. Some scholars even used terms like 'Industrial Revolution of the thirteenth century', an 'early Industrial Revolution', etc., to refer to major changes in that period.[1] However, most studies refer to the developments in the mid-eighteenth century as the Industrial Revolution.

Alvin Toffler, for instance, classified the period up to 1750 as the First Wave – a predominantly agricultural society; 1750 to 1960 as the Second Wave – industrial society; and 1960 onwards as the Third Wave – Information Age. Each Wave was characterised by the transformation in the energy system, production system, and distribution system.[2] Another classification refers to the period from 1750 to the 1960s as the Industrial Age, and the period after that as the Second Machine Age. However, in this chapter, the period from 1750 to the 1830s is referred to as Industry 1.0, the 1850s to 1950s as Industry 2.0, the 1960s to 2000 as Industry 3.0, and from 2000 to the present as Industry 4.0.

The First Industrial Revolution (Industry 1.0)

The Industrial Revolution began in eighteenth-century England and was characterised by the replacement of human skills with mechanical devices, and human and animal power with inanimate power – in particular, steam. The core industries of this revolution were textiles, iron and steel, heavy chemicals, steam engineering, and railway transport.[3] The first ship to use steam power appeared in the UK in 1812, and by the 1860s, virtually all new ships used coal as their source of power. Railway transport started in the north of England in 1826, with Britain playing a leading role in diffusing and financing new technology in the railways and shipbuilding.[4]

The Second Industrial Revolution (Industry 2.0)

The period from 1870 to 1914 is often referred to as Industry 2.0, characterised by the expansion of electricity, petroleum, and steel.[5] During this period, a cheaper method of steel production was invented[6] and steel replaced iron in construction, machines, railroads, and shipbuilding. Electrical generators were introduced, and public electricity replaced candles and lamps, facilitating the invention of the telephone and the design of the light bulb. Electricity began to be used for transportation as radio waves were successfully transmitted across the Atlantic in the early twentieth century. Petroleum became available as fuel and internal combustion engines led to the development of automobiles and aeroplanes.[7] The advent of electricity and assembly lines made mass production possible.[8]

The Third Industrial Revolution (Industry 3.0)

Industry 3.0 began in the 1960s and is characterised by the Digital Revolution. It was built around advances in semiconductors, integrated circuits (IC), computers, internet, and communication technologies. Many innovations in semiconductors helped improve communication and data processing. In 1958, ICs successfully demonstrated their abilities in enabling the production of digital circuits with higher complexity.[9] Since then, the complexity of ICs has approximately doubled every year as per Moore's law.[10] High complexity and condensed electronics fuelled the expansion of computers and other digital applications, including communications, wireless technologies, etc. This enabled the introduction of mainframe computing in the 1960s, personal computing in the 1970s and 1980s, and the internet in the 1990s.[11] Processor speeds exceeding 5GHz, trillion computational operations per second (Teraflops), and bandwidths in excess of 100 Gbps in a single fibre link have facilitated the development of applications across many fields like finance, manufacturing, supply chains, medicine, etc., leading to Industry 4.0.

The Fourth Industrial Revolution (Industry 4.0)

The Digital Revolution of Industry 3.0 triggered a set of applications that are fundamentally different from the previous Industrial Revolutions, as they are a fusion of diverse technologies that interact across the physical, digital, and biological domains.[12] Some scholars prefer to classify the new generation of applications as Industry 4.0, considering the sheer speed of change, diversity across fields, and unprecedented transformation of processes.[13] These technologies have found generic applications in diverse fields and have also had an impact on mercantile marine and warship-building. The Fourth Industrial Revolution facilitates 'Industry 4.0'; more specifically, 'Shipyard 4.0' or 'smart' shipyards. "Uber, the world's largest taxi company, owns no vehicles; Facebook, the world's most popular media owner, creates no content; Alibaba, the most valuable retailer, has no inventory; and Airbnb, the world's largest accommodation provider, owns no real estate."[14]

A major enabler of Industry 4.0 has been successive generations of wireless technologies that facilitated newer applications, making higher speeds of data exchange possible. For example, 1G was analogue cellular; 2G technologies such as CDMA, GSM, and TDMA were the first generation of digital cellular technologies; 3G technologies such as EVDO, HSPA, and UMTS brought speeds from 200 kbps to a few megabits per second; and 4G technologies such as WiMAX and LTE made it possible to scale-up to hundreds of megabits and even reach gigabit-level speeds. 5G technologies usher in three new aspects: greater speeds to move more data, lower latency to improve response time, and the ability to connect more sensors and smart devices simultaneously.[15] 6G technologies, which are already in the works, transcend the spectrum to terahertz and latencies of the order of 1 millisecond (compared to 50 ms of 4G and 5 ms of 5G), enabling the next level of AR/VR and the IoT.[16]

Increased processing powers at higher speeds, greater on-chip complexity, and lower cost are facilitating large numbers of innovations across many fields. For example, the Encyclopaedia of DNA Elements (ENCODE) project – started in 2003 by the National Human Genome Research Institute (NHGRI) in the US – identified functional elements encoded in the human genome sequence, which is a major step in predicting natural and disease-related genetic variation.[17] There are increasing instances where an AI-based system detects cancerous tumours in CT scans, with an accuracy of 95% against human eyes' 65%.[18]

High-speed, secure digital connectivity and advancements in distributed processing have made it possible to connect a number of appliances over the Internet (as in the case of the Internet of Things), or encrypted public ledgers that make all transactions transparent (as in blockchain technology popularised by the digital cash 'Bitcoin'), and many other applications. Entrade[19] is a good example of a remotely controlled clean energy company. The company manufactures micro power plants that generate power and heat, and makes use of various types of readily available waste or biomass. The power plant is a fully automated and remotely managed solution that is housed in a shipping container built on Schneider Electric's EcoStruxure architecture and IoT. Entrade remotely controls and monitors 250 plants in Liverpool, UK, from its headquarters in Germany.[20]

The Onset of the Digital Age: Transition to the Information Age

In the early 1950s, the US had the most extensive and advanced Military Industrial Complex (MIC), and the highest surplus capital for research and international rebuilding assistance. The US focussed on funding research in the areas of science and technology, ensuring technological and military dominance. One of the areas that attracted major funding was research addressing the need for increasingly complex computation and miniaturisation for onboard guidance computers of missiles, and advanced algorithms for interceptions. Research in these areas led to fast evolution in computers, networks, and communications.[21] These technologies also found increased demand in commercial applications like personal computers (PCs), local area networks (LANs), internet, cellular telephony, etc., which were soon being manufactured in Asia by globally integrated enterprises which were mega multi-national companies facilitating the Information Age.[22] Since the countries of maritime Asia were already established manufacturing hubs, many of them could absorb the technological advancements in a short time.

The Information Age required more workers with diverse and continually evolving skills (white-collar workforce). There was, therefore, a reduction in the erstwhile industrial workers who were trained for routine repetitive work and were essentially interchangeable (blue-collar workforce). Societal changes during these transformation periods provided impetus to major educational changes. For the first time in the twentieth century, the demography of the workforce in the US showed an appreciable change, and in 1955, the white-collar jobs of knowledge workers and those in the service sector exceeded the blue-collar ones, primarily in manufacturing.[23] This compares to the demographic change that occurred between 1890 and 1910 when a major proportion of the US workforce shifted from agricultural to industrial production.[24] A few decades later, a similar trend began to be seen in maritime Asian countries.

In addition to the availability of the best infrastructure and capital, two factors contributed immensely to the American lead in innovations in Industry 3.0 technologies. First, the establishment of world-class universities which attracted the best talents from all over the world. These universities were integrated into the MIC for research, incubation, and funding.[25] Second, initiatives by the government to facilitate getting the best scientists to the US and channel their efforts in the required direction. One such example is 'Operation Paperclip', a post-war US intelligence programme.

In May 1945, more than 1600 technologists were brought to the US – from wartime Germany – and employed in research and development in their respective fields of achievement. These fields covered the development of rockets, chemical and biological weapons, aviation and space medicine, and many other armaments. Prominent among those technologists included Otto Ambrose, Theodor Benzinger, Kurt Blome, Walter Dornberger, Siegfried Knemeyer, Walter Schreiber, Walter Schieber, and Wernher von Braun. Many of them rose to receive the US government's highest science achievement awards, and many of them headed prestigious American agencies like NASA in the following years.[26] A similar exercise was also executed by the Soviets to use German talent for their research and development; the Chinese followed suit after the Cold War.

After the disintegration of the Soviet Union in 1989, many scientists – especially those associated with the MIC – found attractive avenues in major emerging economies where they could emigrate and find employment. In late 1994, as many as 1000 Russian specialists were reported to be working in China to improve China's nuclear and rocket programmes.[27] In the US, the Soviet Scientists Immigration Act of 1992 allowed "Soviet" weapons specialists to enter the US permanently without receiving an offer of employment from the US.[28]

Proliferation of Commercial Off-the-Shelf Technologies (COTS)

Initial developments in the field of ICT served to meet the increased computation in compact containers for military applications, although eventually, those that proved to be of practical utility diffused to the consumer sector. The rise in the market size for these products was at a level unprecedented in history. This led to the rise of multinational companies whose reach transcended the boundaries of nation-states and whose net asset value exceeded the annual budget of many nations themselves. The high growth of the market attracted companies' research funding of the order that was difficult to match by even the major powers' defence budgets. Upgrades with stringent specifications were introduced faster in the commercial market, and the designers of defence equipment found it increasingly attractive to use upgraded products as building blocks to enhance capability. This was the initial motivation for the proliferation of commercial off-the-shelf technologies (COTS) in military equipment. COTS technologies are extensively used in marine applications as well as in warships' weapon systems.

One of the most significant examples of the evolution of COTS technologies is that of the GPS. In 1963-64, a US Air Force study on Space Policy and Doctrine (SPAD) concluded that a space-based navigation system could serve a large variety of military and civilian uses. At that time, the US Navy already had an operational satellite system 'TRANSIT' that had been developed by researchers at the Johns Hopkins Laboratory. The TRANSIT system met the requirements of naval ships and submarines but had limitations in accuracy and time required to receive signals from the satellites and calculate position location. Although the Navy developed TIMATION as a follow-on system, the USAF planners jointly developed NAVSAT with the Aerospace Corporation, introducing the concept of using a constellation of four satellites to provide accurate and fast positional information; this was developed further under a government-funded programme '621B'.[29] Programmes TIMATION and 621B were subsequently conjoined, and a GPS was developed using the technologies of the simultaneous use of two frequencies and precision clock requirements. This was funded for the development of dual-use – i.e., for the US defence forces and civilian use worldwide – and was approved by the Pentagon in 1973.[30] NAVSTAR GPS was the first US Department of Defense programme which was directed to be managed as a Joint Service Development Program and evolved as a full-fledged positional information service by the early 1990s. The technology that had initially been conceptualised for ships at sea has since evolved as a stealth utility worldwide, providing positional information across platforms.[31] The selective availability feature through which the US government controlled the availability or degradation of GPS services was removed in 2000, and presently, the new frequency and features enable accuracies of the order of a few millimetres.[32]

Major powers like the US officially enacted the use of COTS in military systems in the mid-1990s.[33] Although defence production accrued savings in cost, governments found it increasingly difficult to exercise control on future requirement specifications, enforce effective technology denial regimes, and prevent non-state actors' access to these technologies. For example, in 2009, militants in Iraq could intercept live video feeds from the US Predator drones, potentially providing them information to evade or monitor US military operations by using a $26 COTS software.[34]

Shift of ICT Manufacturing to Asia

Towards the final decade of the twentieth century, ICT products like Personal Computers, cell phones, networking devices, robotics, CNC machines, etc., commanded a major share of the global market. The core industrial infrastructure and skill sets available in the Asian countries were, by now, capable of absorbing the technology and outperforming the West in terms of production cost. Since profits drive multinational companies, MNCs found ways to leverage economies of scale rather than considering the national interests of their home country. The Original Equipment Manufacturers (OEMs) from the West set up production units in Asia to maximise profits. This further aided the emerging Asian economies' growth – them being populous countries and now experiencing a spurt in middle-class numbers.[35] The emerging Asian consumer surfaced as the new pillar of consumption to propel the global economy. This, in a way, began to diminish US influence, as also its capacity to control the dynamics of the global economy.[36]

The economic and industrial progress after the Second World War also helped the Asian countries increase their military power. This was like the emergence of the new industrial powers in the latter half of the nineteenth century. However, the intrinsic nature of the ICT technologies of Industry 3.0 ushered in an Information Age characterised by seamless connectivity, easy access to knowledge resources, and unprecedented processing power available to a common user. Non-state actors now had potential access to capabilities which were earlier exclusive to governments alone, and this made asymmetric warfare a serious problem. The Information Age also popularised social media, making it a tool for perception management and shaping the battlespace. This ushered an era of 'soft power' – a significant component of military power, and 'smart power' – the ability to combine hard and soft power resources into effective strategies.[37]

Technologies of the Fourth Industrial Revolution (Industry 4.0)

Industry 4.0 began at the turn of the twenty-first century and built on the Digital Revolution. It has arisen from the fusion of a vast spectrum of technologies across physical, digital, and biological domains, and a confluence of emerging breakthroughs in wide-ranging fields such as artificial intelligence (AI), robotics, Internet of Things (IoT), autonomous vehicles, 3D printing, nanotechnology, materials science, energy storage, quantum computing, etc.[38] In the maritime sector,[39] Industry 4.0 technologies have proliferated significantly in applications like autonomous ships and submarines, 3D printed models in design, 3D printed one-of-a-kind parts during construction and spares during repairs, AI- and ML-driven maintenance and support management systems, and blockchain-based transactions between stakeholders in the shipping chain. As these technologies take centre stage and find applications across all sectors, including shipbuilding, the countries in maritime Asia are now competing to lead in the innovation and application of some of these technologies.

Components of Industry 4.0

Artificial Intelligence

Artificial Intelligence (AI) started as the branch of computer science that aimed to make computers follow logical steps to do some of the basic functions that humans performed, using common sense and evolved from imitation, extension, augmentation, and finally reaching human-level AI. In 1956, John McCarthy used the coinage 'Artificial Intelligence' and proposed a summer research project on the subject at Dartmouth College, involving scientists of psychology, mathematics, computer science, and information theory — marking the beginning of AI as the science and engineering of making intelligent machines, as well as recognising it as a field of research.[40] Technologies driving AI have since evolved from conventional programming using logical steps, heuristics using neural networks, and machine learning using big data analytics, to the ability of self-evolution. As a part of AI, machines simulate human intelligence processes like learning (acquiring information and the rules for using that information), reasoning (using rules to reach approximate or definite conclusions), and self-correction to find application in expert systems, speech recognition, and machine vision.[41]

AI and ML have been applied in diverse fields with great success and are transforming even traditional sectors like agriculture. In a vertical bowery farm at the industrial park in New Jersey, all activities – including how much to water each plant, the intensity of light required, when to harvest, etc. – are controlled by an AI-powered algorithm that continuously evolves.[42]

AI-enabled advanced robotics are increasingly used in the workflow of shipbuilding where 'Robotics Process Automation' (RPA) is programmed to perform high volume repeatable tasks. Hyundai Heavy Industries Co. uses a 670-kilogram industrial robot to curve and weld steel plates through remote connectivity between the machine and design software.[43] Artificial intelligence methods are used in the preliminary ship design of cargo ships to optimise performance and achieve minimum propulsion power and maximum deadweight simultaneously.[44] AI- and ML-powered unified applications are built on cloud-based platforms and are increasingly used to augment situational awareness and decision support systems by providing track prediction, considering the ship's hydrodynamic design parameters and anti-collision regulations. These new-generation applications prove useful to the entire marine ecosystem, providing real-time information support by de-risking vessel operation on the bridge of the ship, shore-based fleet operation control, training, and post-voyage analysis. 'A-Suite' is one such application built on the THESIS platform by Transas.[45]

AI's application has added advanced features like in-built diagnostics, fault tolerance, fault indication, and recommendation of repair action in case of catastrophic failure. However, a major evolution has been to progressively reduce the role of humans in achieving designed functionality. This is more pronounced in warships where *advanced weapons and sensors connected on a common grid with sensor-to-shoot capability* and *AI-enabled decision support systems* are now evolving towards a network-based virtual behemoth capable of delivering strategic intent. CANES (Consolidated Afloat Networks and Enterprise Services) is being developed in the US as an AI-powered combat network to seamlessly connect ships, submarines, and Maritime Operations Centres, likely to be implemented to connect 190 entities.[46]

Machine Learning

Machine learning has evolved as a field of study that gives computers the ability to learn without being explicitly programmed. A computer programme, based on an algorithm that computes available data, learns with experience; it executes a task in which its performance is measured and improves with experience. ML originated as one of the fields of AI but subsequently developed as a new capability for computers in industrial applications. There are different types of ML such as 'supervised' – where the algorithm learns based on a pre-defined logic to meet systems requirements or 'un-supervised' – where the algorithm uses an intuitive approach and reviews data to draw inferences, reinforcement learning, and recommender systems.[47]

Supervised ML is increasingly being used to improve the maintenance forecast of running machinery and systems onboard ships, helping to plan work whilst in the harbour so that systems are available at sea. The systems and equipment on board ships (merchant vessels as well as warships) were traditionally maintained as per a planned preventive maintenance (PPM)[48] schedule which was largely time-based. Over the years, this system paved the way for condition-based maintenance (CBM),[49] ensuring the optimisation of maintenance cost and system availability at sea. CBM is now being replaced by predictive maintenance where systems parameters are captured in real-time, and AI-powered advanced analytics and ML tools are used to recommend just-in-time maintenance as well as to avoid failure. The science of maintenance is now graduating towards prescriptive maintenance where the system builds on CBM and is at the intersection of big data, analytics, machine learning, and AI – integrating the functionality of a cognitive system and an asset management system.[50]

ML methods can be used to predict the level of corrosion of a ship by modelling data pertaining to its age, location, structural members, type of coating, etc.; they prove to be a powerful tool to assess the hull state and accordingly ascribe any maintenance action that may be required.[51]

Unsupervised learning – which is also referred to as deep learning – and some other methods are applied in intelligent awareness systems on ships to identify, classify, and track objects that a ship encounters at sea. These systems apply advanced image recognition algorithms and contribute immensely towards the safety of ships at sea and also form a major component of combat management systems on warships. Rolls Royce has signed an agreement with Google to use its Cloud Machine learning engine to train its AI algorithms further, thereby enhancing the capabilities of its intelligent awareness systems to become smarter and adapt self-learning, which can be put to use in autonomous ships.[52]

Virtual Reality/Augmented Virtual Reality

Virtual Reality (VR) and Augmented Reality (AR) are rapidly emerging technologies that are revolutionising the design, maintenance, and training in shipbuilding. VR is an application where a virtual location is digitally created and the physical presence of the viewer is simulated in a manner that the viewer can use headsets, gloves, etc., to move around and interact with the virtual environment. In AR, physical objects are identified in a location and information related to them is augmented by computer-generated inputs using devices such as smartphones, tablets, and head-mounted displays that detect their position relative to the real-world environment around them.[53] Features of AR and VR are also combined in a single application referred to as 'mixed reality'. One such application is being developed for shipbuilding and maintenance by a Norwegian company, using Microsoft HoloLens technology in a headset, thereby making it possible to see reality with an overlay of digital information, move around on the ship, talk on the phone, and cooperate with colleagues sharing the same picture at any remote location.[54]

An early industrial application of AR was in the 'head-up see-through displays' used by Boeing in the 1990s to augment the visual field of the worker with information related to the tasks that he was performing.[55] By the late 1990s, companies like Airbus, EADS, BMW, and Ford had invested in the development of mobile IAR applications.[56] Many AR/VR-enabled software applications are now available as open-source, free, or commercial-off-the-shelf, and are supported by different platforms like Windows, Linux, Android, or iOS. Shipyards choose their preferred software; compatible hardware like tablets, handheld, or head-up displays; advanced game engines, etc. and develop an integrated system which has features of geo-location, localisation, and mapping to map an environment, track movements, and aid navigation.

A common architecture of an integrated enterprise-wide solution for 'Shipyard 4.0' – common to warships and merchant shipbuilding – consists of a Visualisation and Interaction (VIS) layer of handheld and head-mounted devices used by workers onboard the ship, under construction or on the shop floor; a Data Transport System (DTS), which is a combination of wireless and fibre optic networks to transport the data to the Data Abstraction System (DAS) integrated into the cloud with the ERP; third party software; IoT; and mobile edge computing systems. Many shipyards are investing in the development of such VR/AR-intensive technologies to enhance productivity. Navantia shipbuilders created a joint research unit Navantia-UDC (University of A Coruna) at the end of 2015 to study the application of different Industry 4.0 technologies to shipyards, with one of the verticals being 'Plant Information and Augmented Reality'.[57]

3D virtual reality models are replacing erstwhile wood or cardboard mock-ups; stakeholders can now walk through the ship before it even gets off the drawing board to indicate any changes required. Unlike the 2D CAD drawings which require engineering skills and knowledge, the 3D models are used by stakeholders to immerse themselves and contribute towards refining the design, evaluating the piping layout and cable routing, unshipping route, and access for the maintenance of fitted equipment. This obviates a re-work and optimises the overall cost and build period. BAE Systems has used Oculus Rift, a VR headset that presents video or computer graphics in 3D for the video game industry to build 3D full-scale models of the OPVs under construction and to design the Type 26 Frigate for the Royal Navy.[58]

AR is replacing drawings, bringing cameras into the shipyards, and putting computers in the hands of everybody. It builds on the inputs from the 3D models and is extensively used for training, knowledge management, and to add value in construction activities such as welding, painting, quality control, etc. During the construction of very large ships like aircraft carriers, temporary steel is installed at various locations in the initial stage and removed later. Newport News Shipbuilding (NNS) employed AR to locate and remove temporary steel while constructing the USS *Gerald R Ford* by providing step-by-step instructions to workers, using visual overlays which showed temporary steel in a different colour on the handheld devices. NNS started exploring AR in shipbuilding in 2011 and is developing commercial AR applications in collaboration with Index AR Solutions. It aims to make CVN-80, the future USS *Enterprise*, the first paperless aircraft carrier.[59] Such applications are attracting large investments and AR/VR is expected to create a global market of US $80 bn by 2025.[60]

3D Printing: The 'Additive Manufacturing' Process

3D printing or Additive Manufacturing is a method in which an item is manufactured by building up layers of two dimensions (2D), repeatedly, to add up to the final product. The basic requirement of this process is to prepare a three-dimensional (3D) digital model of the product and then extrude specified material in a controlled environment to mould up, layer-by-layer, to manufacture the desired product. 3D printing was initially used for 'Rapid Prototyping'.[61] This manufacturing process has rapidly grown to connect the digital and real world and is now being used to manufacture the product itself.

Charles Hull made the Stereo-lithography Apparatus (SLA) machine in 1983 and subsequently, other technologies emerged like Selective Laser Sintering (SLS), Fused Deposition Modelling (FDM), Direct Metal Laser Sintering (DMLS), etc.[62] Prototyping is still the largest application of 3D printing and is also used in tooling and casting. The added capabilities of this technology for customisation and personalisation make it most suitable for application in fields including medicine, aerospace, automotive, jewellery, art, sculpture, the food industry, and shipbuilding. Ushering a connection between the virtual and real world, advances in digital modelling were integrated to the capability to extrude the desired material in a controlled environment to form layers in two dimensions and manufacture. Organovo, a San Diego-based company, has developed a bioprinting process that takes cells from donor organs and turns them into printable bio-ink. Layers of this ink have been successfully used to build small sections of liver tissue.[63]

The shipbuilding industry uses 3D printers to print the models of ships during design. Unlike other industries, the printed models are used as functional models and subjected to mechanical loads for optimising the design, especially of appendages like rudders. A German hydrodynamics research organisation – Hamburg Ship Model Basin (HSVA) – reported a 70 per cent slash in lead production times, and an overall cost reduction of around 30 per cent after using 3D printers for two years.[64] The use of 3D printed models makes it possible to have more iterations for the early finalisation of design, and thereon present the promotional models for the customer's better appreciation. This technology complements lean and 'just in time' processes, as well as the one-piece flow in naval and maritime constructions. 3D printing is being increasingly used to manufacture micro panel profiles, non-standard brackets, and outfitting equipment like double bottoms, side shells – single and double skins, decks, and longitudinal bulkheads. A major advantage is that it provides the capability to manufacture 'one-of-a-kind' steel parts, or smaller and complex parts without having to machine it or outsource it to a third party, thus substantially reducing the cost.

A major impact of 3D printing is in the refit and repairs of ships and the logistic support. This applies to both the warship as well as the merchant marine segments. Ships, by their very nature, have longer life cycles compared to the rapidly evolving technologies in the commercial space. Accordingly, thorough life cycle logistics support often calls for complex supply chain management and large and expensive inventory costs. The issue is also complicated by the urgency of demands, in case of breakdowns and failures, as well as the geographic dispersion of the affected ship. 3D printing offers a great solution by way of capability: to 'print' a spare on demand and just in time. A palm-sized drain strainer orifice was 3D printed by Newport News shipbuilders, using a highly digitised process that deposited metal powder, layer-by-layer, to the 3D marine alloy part, installed on a steam line onboard the aircraft carrier USS *Harry S Truman*. A 3D printing lab is also being installed onboard the USS *George Washington* during its ongoing mid-life refuelling and complex overhaul.[65] Items that are already being printed include gaskets, 'O' rings, small parts of specific shapes and dimensions, special tools required, valves of different sizes and materials, etc. The advantage accrued is not only in terms of prompt repairs ashore, but also freeing up space, since ships may have no necessity to stock onboard spares. Whilst newer warships are already being equipped with 3D printers, merchant marine ships are utilising 3D printers installed in their next ports of call for delivery of the spares, which can be printed on demand from sea. Options being exercised by each segment are purely on operational and commercial considerations.

As 3D technology matures further and comes closer to the mass production techniques of today, more and more goods will be manufactured at or close to their point of purchase or consumption.[66] Fewer finished products, therefore, will be required to be shipped from across the globe. This may, in the distant future, mean that the bulk of shipping cargo will consist of raw materials and 3D print cartridges.[67]

Blockchain

The global financial crisis of 2008 was caused by the deregulation of the financial sector and fuelled the idea of an entirely new form of currency which could be transparent and global. A White Paper published under the pseudonym of Satoshi Nakamoto in October 2008 presented a blueprint of a peer-to-peer electronic cash system.[68] The blockchain-based digital encrypted currency with a distributed ledger system was first released in January 2009 and was managed by bitcoin.org.[69] Although the overall value of the global cryptocurrency market is presently estimated to be over $700 billion,[70] the underlying blockchain technology has been employed in diverse fields, including shipbuilding.

Shipping is a complex supply chain, and requires a trail of contracts, payments, approvals, insurance, and tracking, as the cargo moves from the originator to the destination. Delays in any of the intermediate activities affects the efficiency of the entire supply chain and translates to delays and higher cost. Blockchain technology offers high efficiency, security and transparency in the form of a digital ledger system that makes it the perfect solution to obviate paper trails. Container shipping lines are teaming up with technology companies to work out a protocol to integrate manufacturers, banks, insurers, brokers, and port authorities around the world over a blockchain-based system. 'TradeLens' is a blockchain-based shipping platform created by Maersk and IBM to improve the efficiency and security of the global supply chain. It uses blockchain as a digital supply chain foundation, establishing a single shared view of a transaction while keeping privacy and confidentiality secure for shippers, shipping lines, freight forwarders, port and terminal operators, inland transportation, and customs authorities.[71] Once fully implemented, this could be the biggest revolution in shipping after the 1960s introduced fixed-sized containers; it is estimated to generate an additional $1 trillion in global trade, according to the World Economic Forum.[72]

In the shipbuilding industry, blockchain technology can integrate the entire supply chain of designers, financers, sub-contractors of services, suppliers of equipment and construction materials, quality control inspectors, auditors, and customers. Transparent and secure transactions inherent to blockchain technology will ensure that the multiple agencies involved at each stage of construction are on the same page, thereby securing trust and removing bottlenecks caused by communication gaps. Maritime Digital Supply Space (MDSS) is a joint project started in 2017 by Lappeenranta University of Technology (LUT), Finland, to develop a blockchain-based solution for shipbuilding, involving academia, shipyards, and industries like Rolls Royce and Wartsila.[73]

A blockchain has the inherent feature of error detection by using consensus and ensures data fidelity by making any valid operation immutable. These two features address the lacunas in the Automatic Identification System (AIS) – widely used by the maritime world – and can be addressed by using blockchain technology. The other potential application of blockchain technology under consideration is tactical networks, where issues of ambiguity and manipulation of target tracks by sources external to a coalition can be addressed using blockchain technology to provide a secure and robust Recognized Maritime Picture (RMP).[74]

Autonomous Ships

Autonomous ships are platforms that operate independently under the control of an onboard decision support system with a provision of control from a remote station. These have a mix of remote and automated technologies, set to evolve from 'unmanned remotely controlled' through 'partly automated' to fully 'autonomous' ships. A fully autonomous vessel now will not need to have spaces and systems required for human habitability, and can also cater for additional cargo space. Advancements in the fields of AI and robotics could make it possible to man autonomous ships by robots and eliminate damages due to human errors which presently account for more than 75% of all accidents in the shipping sector.[75]

At least four companies are investing in the development of autonomous ships – Rolls Royce Holdings Plc, BHP Billiton Ltd, Tokyo-based fertiliser producer Nippon Yusen KK, and YARA International ASA.[76] Rolls Royce Marines is leading a group of ship designers, equipment makers, and universities in a €6.6 million Advanced Autonomous Waterborne Application (AAWA) initiative funded by a Finnish funding agency, which had aimed to prove autonomous technology for short coastal runs by 2020 and ocean-going by 2025. Similar initiatives by Kongsberg and Yara were ongoing in the 2019-2020 timeframe; a Japanese consortium is looking to operate a fully autonomous ship by 2025.[77] DNV GL – supported by Transnova, Norway – has also developed an innovative shipping concept as a solution in the short-sea segment. Named 'ReVolt', this vessel is 60 metres in length, has a maximum speed of 6 Knots, is battery operated, has a range of 100 nautical miles, and has a cargo capacity of 100 twenty-foot containers. ReVolt is autonomous and requires no crew.[78]

Autonomous warships and submarines are being developed at a rapid pace by most of the major navies of the world. Sea Hunter is a technology demonstration vessel under the Anti-Submarine Warfare (ASW) Continuous Trail Unmanned Vessel (ACTUV) programme of DARPA in the US. It has successfully completed sea trials and could transition to the US Navy as a Medium Displacement Unmanned Surface Vessel (MDUSV).[79] Large Displacement Unmanned Underwater Vehicles (LDUUV) and Extra Large Unmanned Undersea Vehicles (XLUUV) are also being developed involving the Navy, Boeing, Lockheed Martin, and Huntington Ingalls Industries.[80] The (Chinese) People's Liberation Army Navy (PLAN) is also likely to acquire an AI-enabled XLUUV, which is being developed by the Shenyang Institute of Automation under the Chinese Academy of Sciences (CAS).[81]

Plymouth University, autonomous crafts specialists MSubs, and Shuttleworth Design are partnering on a project to design, build, and sail the world's first full-sized, fully autonomous unmanned ship across the Atlantic Ocean. The Mayflower Autonomous Research Ship (MARS) has an overall length of 100 feet and uses state-of-the-art wind and solar technology for propulsion. After a year of the testing phase, the Atlantic crossing marked the 400th anniversary of the original Mayflower sailings from Plymouth, England to Plymouth, Massachusetts, USA.[82] The ship sailed from Plymouth, UK on 15 June 2021. Growth in the processing capabilities and integration of onboard sensor data have made it possible to graduate from unmanned remotely-controlled vehicles to partly automated and finally to autonomous vehicles in air, land, and sea (surface as well as sub-surface). On-demand public shuttle service of electric-powered autonomous mini-buses is likely to be available in

Sentosa Islands and some other locations in Singapore for which trials are in progress.[83]

Quantum Computing

A quantum computer uses quantum particles,[84] placed in a highly controlled environment to form a 'qubit', similar to the binary digit (bit) of a conventional digital computer. Quantum particles have inherent properties of superposition,[85] entanglement,[86] and interference,[87] which are utilised to solve a problem. A qubit is made by cooling a metal such as Niobium to extremely low temperatures (of the order of 10-15 milli-Kelvin using liquid helium), in a controlled magnetic environment, in which it is a superconductor. Under these conditions, a quant particle like an electron or a nucleus is isolated, with its magnetic spin state being maintained or controlled – pointing 'up' or 'down' to be read as 'one' or 'zero', respectively. Multiple qubits can be connected using a superconducting loop, and the state of a qubit is controlled using microwave signals. Whilst many quantum computers utilising 50 to 72 qubits are now available, the research effort is focussed on achieving higher qubit systems – Quantum Error Correction (QEC) and higher fault tolerant systems.[88]

In March 2018, Google unveiled a 72-Qbit quantum processor named Bristlecone at the annual American Physical Society meeting in Los Angeles. Bristlecone is envisaged to demonstrate Quantum Supremacy,[89] investigate first- and second-order error correction using surface code, and facilitate quantum algorithm development on actual hardware.[90] Chinese scientists have found new properties of a particle – Majorana Fermion – which can be used to make the next generation of more fault-proof quantum computers.[91]

Many of the natural phenomena and chemical reactions are inherently quantum mechanical in nature. Since qubits follow the same natural laws, the use of quantum computers can lead to the discovery of new materials and medicines, and also find applications like Shor's algorithm for factorisation and Grover's algorithm for unstructured search.[92] Some of the progress in fields like genomics – as in the creation of the Cancer Genome Atlas; material science – as in the Materials Genome Initiative; advanced battery chemistries to drive clean energy. Economy and manufacturing Industry 4.0 demands higher computational speed which could be met by quantum computers of the future. The requirement of increasing transistor density on a silicon wafer is becoming more difficult due to the limits of technological boundaries that are challenging growth rates dictated by Moore's Law. An alternate process is emerging, wherein the system learns how to use bits to drive atoms, which could involve advanced machine learning algorithms and new computing architectures, such as quantum computing and neuromorphic chips.[93]

Advanced Materials

Advanced materials are lightweight materials developed from compounds at a molecular level, and are broadly grouped under three categories, viz. composites, polymers, and nanomaterials, to reduce the overall weight while maintaining or enhancing performance.[94] These materials are generally lighter, stronger, recyclable, and adaptive. There are now applications for smart materials that are self-healing or self-cleaning, metals with memory that revert to their original shapes, ceramics and crystals that turn pressure into energy, etc.[95]

Materials that are based on composite carbon fibre and fibre-reinforced plastic are being used in new construction ships because they are lighter and can reduce fuel consumption and air pollution. Composite carbon fibre has been used to build the superstructure of warships such as the *Zumwalt*-class destroyers of the US Navy, the *Steregushchiy*-class corvettes and stealth frigate *Admiral Gorshkov* of the Russian Navy, and corvettes for the Indian Navy.[96] Advanced nanomaterials such as graphene, which is 200 times stronger than steel, a million times thinner than human hair, and also a good conductor of heat and electricity is being used in a wide range of applications – from mobile generators that are powered by hydrogen, potable water generation through seawater filtration, a potential replacement for silicon in electronics, to IR-sensing wearables. Graphene-based paints are virtually impermeable and can prevent corrosion and rust. These hold great potential for application in shipbuilding.[97]

An 'omniphobic' coating material has been developed at the University of Michigan – based on the mathematical modelling and analysis of the molecular structures of all known chemical substances – to predict their behaviour once they are blended. This material is clear, durable and can repel water, oil, alcohol, etc., and could substantially reduce the underwater friction drag of ships, submarines, and Unmanned Underwater Vessels. The Office of Naval Research (ONR) in the US has sponsored the development of this special material to save on fuel expenditures, given that about 80 per cent of fuel is consumed at lower speeds and 40 to 50 per cent of fuel is consumed at higher speeds, in order to overcome the drag due to underwater friction.[98]

Digital Twinning

The digital twin of a device is a virtual representation of elements and dynamics of the device through its design, build, and operation phases, providing insight into the actual environment in which the device operates through its life cycle. Companies such as IBM and GE are working on the Internet of Things as a framework for connecting equipment sensors, data, and models. The main parts of a digital twin are: a physical product in the real space; a virtual product in the virtual space; and connections of data and information that tie the virtual and real products together.[99]

A ship's digital twin holds great potential in providing an enormous amount of data to improve accuracy during the design and build phases, predictive and prescriptive maintenance during the operational phase, and in the future, it will be useful for remote monitoring by expert systems when autonomous ships get deployed. Shipbuilders, systems and equipment designers, manufacturing industries, and research agencies have joined the Open Simulation Platform (OSP) which will develop an open-source digital platform for creating digital twins of ships in order to optimise design, production, maintenance, and sustainability throughout their life cycle. Rolls Royce Marine, the Norwegian University of Technology Science (NTNU), Hyundai Heavy Industries, DNV GL, and Kongsberg Digital are some of the major players, and the OSP is presently running a prototype digital twin of a vessel and a Dynamic Positioning (DP) system conducting a dynamic positioning operation.[100]

Shipyards building warships such as Spain's Navantia shipyard and the US' Newport News are also adopting digital techniques in all spheres – from supply chain management to design and build. Modern warships have an extremely high degree of complexity in design to meet their envisaged roles, and since technology is changing at an ever-fast pace, there is an increased demand for longer ship lives and major mid-life upgrades. The digital twinning technology is envisaged to make the greatest impact in this sphere, since a digital twin will make it possible to visualise the complete impact of the upgrade, and also study the impact on the ship when deployed for the new role envisaged.[101] The digital twin of a warship can be used as a reference system and any deviations in the performance parameters can be shared with the ship to facilitate predictive maintenance and reduce cost. General Electric has proposed to make the digital twin of two bulk freighters – from the Military Sealift Command (MSC) of the US Navy – which are deployed for replenishment deliveries. The ships' actual data will be analysed by an Asset Performance Management (APM) software suite of GE, and will be compared with the digital twin that depicts the propulsion of the ships, the frequencies being used, diesel-operated machines, etc.[102]

The Internet of Things

The Internet of Things (IoT) is a network of physical objects that are embedded with electronics, software, sensors, and connectivity in which various sensors and actuators are interfaced with computers on the network. IoT has been widely adopted in many fields, such as connected cars and homes, smart factories, wearable devices, intelligent infrastructure, and manufacturing industries.[103] New applications are transforming common objects into connected devices over the internet, making ubiquitous computing[104] a reality which grows beyond consumer applications to provide a framework for industries in the form of Industrial IoT.[105] Internet of Everything (IoE) is the networked connection of people, processes, data, and the IoT.

Shipyards are making smart ships by fitting hull monitors, equipment sensors, machinery diagnostics devices, and CCTVs during the construction, and integrating them with compatible software platforms provided by global mobile satellite operators to implement IoT for real-time expert systems' support from remote locations. Inmarsat and Samsung Heavy Industries (SHI) of South Korea have concluded a strategic agreement under which SHI is fitting Inmarsat-approved hardware, which can then be used with Inmarsat's FleetXpress platform through a dedicated bandwidth provided to certified owners.[106] Large volumes of data available from ships at sea are transforming maintenance as well as aiding in the accurate planning of upcoming sorties. These are also the building blocks towards the evolution of autonomous ships.

As more secure and faster wireless digital connectivity over distances becomes available, IoE is emerging as the primary platform for network-centric operations for navies. IoE solutions enable predictable system performance, enhanced situational awareness, and better information management to aid faster decision-making. IoE enhances the efficiency of logistic chains, intelligence, surveillance, and reconnaissance, thereby improving the operational effectiveness of the fleet. A White Paper published by Cisco Systems Inc USA outlined the possibilities for IoE applications in the US Navy.[107] Present-generation warships are already equipped with sensors on propulsion machinery, auxiliaries and weapons, and Fire Control Systems, but as IoE-enabled sensors are getting popular and are available off-the-shelf at lower prices, system designs are now evolving on the IoE backbone. An IoT-based Integrated Platform Management System by Siemens has been installed on two warships of the New Zealand Navy to achieve unmanned machinery spaces in these warships.[108]

Cyber-Physical Systems

A real-world or a physical system that has an embedded computer, integrated with a communication network, and can govern actuators and receive inputs from sensors is broadly defined as a Cyber-Physical System (CPS). Such systems effectively form a smart control loop capable of adaptation and autonomy.[109] A system that combines the capabilities of computing, communications, and data storage in order to monitor or control the entities that exist in the physical world is referred to as CPS.[110]

Modern ships are becoming highly automated, and systems designers are using CPS as building blocks in many onboard systems, such as the Dynamic Positioning System (DPS); emergency shutdown and blow-out prevention system; machinery monitoring and control systems; propulsion system; navigation system; Combat Management System; and fire control and missile systems. Most of these systems have evolved to provide remote monitoring as well as control. One of the most advanced features is collaborative engagement by warships with sensor-to-shoot capability. All these systems already have a high degree of autonomy built in and extensively deploy CPS. Future ships will support such a high degree of ubiquitous computing that the entire ship could be considered a CPS.[111]

The concept of CPS has the potential to transform the production logistics in Shipyards 4.0 by integrating CPS into products, parts, and logistics resources. The implementation of CPS can facilitate demand-oriented production supply, like in Milk Run 4.0,[112] and holistically synchronise materials and information flows. It can also automate the Kanban approach[113] of the traditional mass-production environments to achieve much higher efficiencies. Logistics processes can be optimised by using CPS-based solutions like tracking and tracing heavy load carriers in harbours with RFID positioning systems, complimentary inventory strategies like expediting the required carrier on request, and optimising the traffic flow and magnetic traverse of steel products.[114]

The Post-COVID-19 World and the Fourth Industrial Revolution (Industry 4.0)

Major disruptions of the global supply chains during the COVID-19 pandemic have drawn the attention of the world towards the risks of monopolising sources of supply of critical raw materials for manufacturing as well as finished products. There is a clear global shift towards diversified manufacturing and supply chains.[115] There are major initiatives being taken in the West to facilitate a return of manufacturing to the West, and certainly out of single vendor dominance of specific countries in critical technologies. One such example is CHIPS (Creating Helpful Incentives to Produce Semiconductors) and the Science Act in the US. This initiative aims to provide substantial subsidies and financial resources to bring back the manufacturing of semiconductor fabs to the US.[116] There has also been an increased effort to diversify markets from within the Indo-Pacific countries, to enable more manufacturing hubs to come up and also find innovative solutions to automate and localise manufacturing, using the technologies of Industry 4.0. For example, 3D-printed drones can be produced anywhere in the world, virtually at the same cost if the raw material required is available. Similarly, AI-enabled processes can obviate the use of manpower and intelligent systems can increasingly facilitate the reverse shifting of manufacturing. An interesting aspect of this development is that by now, the capabilities already in place in the Indo-Pacific are so advanced that they are also contemporary competitors. Hence, in many sectors, some of the countries in the Indo-Pacific are in the lead. One of the facets of developments in cutting-edge technologies is that they work well in collaboration rather than in isolation. Fields like AI and ML intricately relate to the user database, and more is better. Data is increasingly becoming the new 'oil', and China is becoming the new Saudi Arabia.

The Fourth Industrial Revolution (Industry 4.0): Security and Regulatory Imperatives

Technologies under Industry 4.0 extensively use digitisation of existing phenomena. Any phenomenon that gets digitised experiences rapid growth in the steps of the 'Six Ds of exponentials: digitalisation, deception, disruption, demonetisation, dematerialisation and democratisation'.[117] Once digitised, the growth pattern of that product begins to follow a trend similar to that of information technology, which has largely grown as per Moore's law. An indicator of the impact of the digitised world's expansion is that the processing power has increased 1-trillion-fold from 1956 to 2015.[118]

Since most emerging applications have a high content of digitisation and networking, the issue of cybersecurity is becoming central to providing assured service. During the Information Age in Industry 3.0, the cyber security industry focussed on protecting the confidentiality of data of an individual. However, Industry 4.0 technologies are getting more integral to human functioning, from health to artificial intelligence and autonomous operations, thus transcending the scope of cyber security from mere data protection to ensuring digital integrity and availability.[119] There is also a requirement to evolve a new regulatory framework that addresses the impact of these technologies. The issue of the legal framework required for applications such as autonomous vehicles is being addressed by international organisations. The International Maritime Organisation (IMO) has designated autonomous and semi-autonomous ships as Maritime Autonomous Surface Ships (MASS) and has commenced work on evolving suitable regulations to be applicable in the future.[120]

Technologies like AI/ML, IoT, 3D printing, autonomous vehicles, etc., are being extensively applied in various sectors including shipbuilding. Since some of the countries in maritime Asia are the world's leading shipbuilding nations, their shipyards are leading in the implementation of Industry 4.0 technologies. Therefore, applications like 3D printing, the Internet of Things, digital twinning, etc., in shipbuilding by the lead shipyards in maritime Asian countries further consolidate their lead position, thereby increasing their profits and strengthening the Indo-Pacific construct.

The overall influence of history, economics, technology, and strategy and the developments in Industry 4.0 technologies in some of the maritime Asian countries, including Japan, South Korea, China, and India will be studied next. This will help in understanding the contribution of shipbuilding in the overall growth of these countries after World War II and the consequential evolution of the Indo-Pacific construct.

NOTES

1   Landes, *The Unbound Prometheus,* ch.3, 1.
2   Alvin Toffler, *The Third Wave,* April 1981 (New York: Bantam Books, 1981), 27.
3   Landes, *The Unbound Prometheus,* ch.3, 1-4.
4   Maddison, *The World Economy,* ch.2, 98.
5   Patricia Chappine, "The Second Industrial Revolution: Timeline & Inventions," *Study.com,* http://study.com/academy/lesson/the-second-industrial-revolution-timeline-inventions.html, accessed 12 September 2018.
6   Landes, *The Unbound Prometheus,* ch.3, 259.
7   Chappine, *The Second Industrial Revolution,* ch.4.
8   Klaus Schwab, *The Fourth Industrial Revolution* (New York: Portfolio Penguin, 2017), 7.
9   Lidia Lukasiak and Andrzej Jakubowski, "History of Semiconductors," *Journal of Telecommunications and Information Technology* vol. 1, no. 1(2010), 3- 9.
10 G E Moore, "Progress in Digital Integrated Electronics," *IEEE Solid-State Circuits Society Newsletter,* vol. 11, no. 3 (September 2006), 36–37.
11 Schwab, *The Fourth Industrial Revolution,* ch.1, 7.
12 Schwab, *The Fourth Industrial Revolution,* 8.
13 Schwab, *The Fourth Industrial Revolution,* 3.
14 Tom Goodwin, "The Battle Is for The Customer Interface," *TechCrunch,* http://social.techcrunch.com/2015/03/03/in-the-age-of-disintermediation-the-battle-is-all-for-the-customer-interface/, accessed 17 September 2018.
15 Sascha Segan, "What Is 5G?", *PCMag India,* https://in.pcmag.com/cell-phone-service-providers/104415/what-is-5g, accessed 16 April 2019.
16 "6G Mobile Network," https://www.rantcell.com/how-is-6g-mobile-network-different-from-5g.html, accessed 31 July 2021.
17 Kelly A Frazer, "Decoding the Human Genome," *Genome Research,* vol. 22, no. 9 (September 2012), 1599–1601.
18 Press Trust of India, "AI System Can Detect Cancer Tumours in CT Scans Better than Humans," *Business Standard India,* https://www.business-standard.com/article/technology/ai-system-can-detect-cancer-tumours-in-ct-scans-better-than-humans-118082500154_1.html, accessed 25 August 2018.

19   "Entrade, the Clean Energy Company," //www.schneider-electric.co.in/en/work/ campaign/life-is-on/case-study/entrade.jsp, accessed 24 September 2018.

20   "Entrade, the Clean Energy Company," //www.schneider-electric.co.in/en/work/ campaign/life-is-on/case-study/entrade.jsp, accessed 24 September 2018.

21   Stefan Thomas Possony, JE Pournelle, and Francis X Kane, *The Strategy of Technology* (Electronic Edition: WebWrights, 1997), https://www.jerrypournelle.com/slowchange/ Strat.html, ch.1, 15.

22   Samuel J Palmisano, "The Globally Integrated Enterprise," *Foreign Affairs* vol. 85, no. 3 (2006), 127, https://doi.org/10.2307/20031973, accessed 18 August 2018.

23   Toffler, *The Third Wave*, ch.1, 138-139.

24   Glenn E Baker, Richard A. Boser, and Daniel L. Householder, "Coping at the Crossroads: Societal and Educational Transformation in the United States," *Journal of Technology Education* vol. 4, no. 1 (1992).

25   Philip G Altbach, "The Past, Present, and Future of the Research University," *Economic and Political Weekly* vol. 46, no. 16 (April 16, 2011), 67.

26   Annie Jacobson, *Operation Paperclip: The Secret Intelligence Program That Brought Nazi Scientists to America* (New York: Brown and Company, 2014), ix-xi.

27   RA Moody, "Reexamining Brain Drain from the Former Soviet Union," *The Nonproliferation Review* vol. 3, no. 3 (September 1996), 94, https://doi.org/10.1080/ 10736709608436643, accessed 13 May 2019.

28   Vladimir Shkolnikov, "Potential Energy: Emergent Emigration of Highly Qualified Manpower from the Former Soviet Union" (Pardee RAND Graduate School, 1994), xx, https://www.rand.org/pubs/rgs_dissertations/RGSD113.html, accessed 13 May 2019.

29   Possony et al, *The Strategy of Technology*, ch.1, 42.

30   BW Parkinson and ST Powers, "Part 1: The Origins of GPS, and the Pioneers Who Launched the System," *GPS World*, 02 May 2010, https://www.gpsworld.com/origins-gps-part-1/, accessed 15 May 2019.

31   BW Parkinson and ST Powers, "Part 2: The Origins of GPS, Fighting to Survive," *GPS World*, 01 June 2010, https://www.gpsworld.com/origins-gps-part-2-fighting-survive/, accessed 15 May 2019.

32   "GPS.Gov: GPS Accuracy," https://www.gps.gov/systems/gps/performance/accuracy/, accessed 17 May 2019.

33   Jacques S. Gansler and William Lucyshyn, "Commercial-Off-the-Shelf (COTS): Doing It Right," *University of Maryland*, September 2008, www.dtic.mil/dtic/tr/fulltext/u2/ a494143.pdf, accessed 15 August 2018.

34   Siobhan Gorman, YJ Dreazen, and August Cole, "Insurgents Hack U.S. Drones," *Wall Street Journal*, 18 December 2009, http://www.wsj.com/articles/SB12610224788 9095011, accessed 15 August 2018.

35   According to a 2017 study by the Brookings Institution, 88 per cent of the next one billion people to enter the middle class, globally, will be Asians.

36   Michael Schuman, "More and More Families Are Joining the Global Middle Class," *US News & World Report*, https://www.usnews.com/news/best-countries/articles/2018-01-23/asian-consumers-becoming-most-powerful-economic-force-in-world, accessed 10 April 2018.

37   Joseph S Nye, Jr, *The Future of Power*, First (United States: Public Affairs, 2011), 23.

38 Klaus Schwab, "The Fourth Industrial Revolution: What It Means and How to Respond," *World Economic Forum*, https://www.weforum.org/agenda/2016/01/the-fourth-industrial-revolution-what-it-means-and-how-to-respond/, accessed 01 August 2018.

39 Refers to shipping, merchant shipbuilding, warship-building, offshore platforms and maintenance and repairs of all these assets.

40 Zhongzhi Shi, *Advanced Artificial Intelligence* (Singapore: World Scientific Publishing Co Pte Ltd, 2014), 1–2.

41 Margaret Rouse, "What Is AI (Artificial Intelligence)? - Definition from WhatIs.Com," *SearchEnterpriseAI,* https://searchenterpriseai.techtarget.com/definition/AI-Artificial-Intelligence, accessed 29 September 2018.

42 Bloomberg, "At This Farm, the Boss Is an AI-Powered Algorithm," *The Economic Times*, 21 September 2018, https://economictimes.indiatimes.com/small-biz/startups/newsbuzz/at-this-farm-the-boss-is-an-ai-powered-algorithm/articleshow/65895095.cms, accessed 24 September 2018.

43 Kyunghee Park, "Labour-Intensive Shipbuilding Industry Employing Robots to Cut Costs," *Business Standard India*, https://www.business-standard.com/article/international/labour-intensive-shipbuilding-industry-employing-robots-to-cut-costs-118041600155_1.html, accessed 16 April 2018.

44 Tomasz Abramowski, "Application of Artificial Intelligence Methods to Preliminary Design of Ships and Ship Performance Optimization," *Naval Engineers Journal,* vol. 125 (September, 2013), 101–12.

45 Sarah Carter, "AI-Powered Maritime Applications Launched by Transas," *ShipInsight*, https://shipinsight.com/articles/ai-powered-maritime-applications-launched-transas, accessed 31 January 2018.

46 "The CANES Evolution," https://www.public.navy.mil/spawar/PEOC4I/Pages/CANESTest.aspx, accessed 03 October 2018.

47 Sumit Das et al, "Applications of Artificial Intelligence in Machine Learning: Review and Prospect," *International Journal of Computer Applications*, vol. 115, no. 9 (2015).

48 A method of carrying out preventive maintenance of equipment as per a planned schedule.

49 A method of carrying out maintenance only when the monitored parameters indicate necessity.

50 Matt Bellias, "The Evolution of Maintenance towards Prescriptive," *Maintenance evolution prescriptive*, https://www.ibm.com/blogs/internet-of-things/maintenance-evolution-prescriptive/, accessed 10 February 2018.

51 Abhishek Chauhan et al, "A Machine Learning-Based Approach to Predict Corrosion Allowance for Ships", *The 28th International Ocean and Polar Engineering Conference, International Society of Offshore and Polar Engineers, 2018*, https://www.onepetro.org/conference-paper/ISOPE-I-18-442, accessed 23 October 2018.

52 Bernard Marr, "Rolls-Royce and Google Partner to Create Smarter, Autonomous Ships Based on AI and Machine Learning," *Forbes,* https://www.forbes.com/sites/bernardmarr/2017/10/23/rolls-royce-and-google-partner-to-create-smarter-autonomous-ships-based-on-ai-and-machine-learning/, accessed 23 October 2018.

53 Bjorn Mes, "Virtual and Augmented Reality in Shipbuilding", *Damen Magazine*, https://magazine.damen.com/innovation/virtual-and-augmented-reality-in-shipbuilding/, accessed 05 October 2018.

54 Paul Bartlett, "Mixed Reality Technology to 'Disrupt' Shipbuilding and Maintenance," *Seatrade Maritime News,* http://www.seatrade-maritime.com/news/europe/mixed-reality-technology-to-disrupt-shipbuilding-and-maintenance.html, accessed 05 October 2018.

55 TP Caudell and DW Mizell, "Augmented Reality: An Application of Heads-up Display Technology to Manual Manufacturing Processes," *Proceedings of the Twenty-Fifth Hawaii International Conference on System Sciences,* vol. 2 (1992), 659–669.

56 N Navab, "Developing Killer Apps for Industrial Augmented Reality," *IEEE Computer Graphics and Applications,* vol. 24, no. 3 (May 2004), 16–20.

57 P Fraga-Lamas et al, "A Review on Industrial Augmented Reality Systems for the Industry 4.0 Shipyard," *IEEE Access,* vol. 6 (February 2018), 13358–75.

58 Alan Tovey, "Virtual Reality Warships: Why BAE Is Diving into 3D," *The Telegraph,* 05 November 2015, https://www.telegraph.co.uk/finance/newsbysector/industry/defence/11210224/Virtual-reality-warships-Why-BAE-is-diving-into-3D.html, accessed 09 October 2019.

59 "Digital Reality Technology: Empowering Today's Knowledge Workers," *Index AR Solutions,* https://www.indexarsolutions.com/digital-reality-technology-empowers-knowledge-workers/, accessed 02 August 2018.

60 Global Investment Research, *Virtual & Augmented Reality - Understanding the Race for the Next Computing Platform,* (Goldman Sachs, 13 January 2016), https://www.goldmansachs.com/insights/pages/technology-driving-innovation-folder/virtual-and-augmented-reality/report.pdf, accessed 10 August 2018

61 A method where design is transformed into a prototype for quick appreciation and feedback from the user.

62 JO Milewski, *Additive Manufacturing of Metals: From Fundamental Technology to Rocket Nozzles, Medical Implants, and Custom Jewellery* (Switzerland: Springer International, 2017), 14.

63 Hasan Chowdhury, "Liver Success Holds Promise of 3D Organ Printing," *Financial Times,* https://www.ft.com/content/67e3ab88-f56f-11e7-a4c9-bbdefa4f210b, accessed 05 March 2018.

64 Eva Grey, "3D Printing: Rising to the Challenge in Ship Design," *Ship Technology,* http://www.ship-technology.com/features/feature3d-printing-rising-to-the-challenge-in-ship-design-4672912/, accessed 08 March 2016.

65 Ben Werner and Megan Eckstein, "Palm-Sized 3D-Printed Part Represents Leap Forward In Shipbuilding," *USNI News,* https://news.usni.org/2018/10/12/palm-sized-part-represents-leap-forward-in-shipbuilding, accessed 12 October 2018.

66 RA D'Aveni, "3-D Printing Will Change the World," *Harvard Business Review,* https://hbr.org/2013/03/3-d-printing-will-change-the-world, accessed 30 July 2016.

67 Tatjana Schork, "Six Theories about How 3D Printing Will Change Logistics", *AEB,* https://www.aeb.com/sg-en/magazine/articles/white-paper-3d-printing.php, accessed on 13 October 2018.

68 Satoshi Nakamoto, "Bitcoin: A Peer-to-Peer Electronic Cash System," https://bitcoin.org/bitcoin.pdf, accessed 20 September 2018.

69 Benjamin Wallace, "The Rise and Fall of Bitcoin," *Wired,* https://www.wired.com/2011/11/mf-bitcoin, accessed 02 January 2021.

70 Will Martin, "The Global Cryptocurrency Market Hit a New Record High above $700

Billion," *Business Insider*, https://www.businessinsider.com/bitcoin-price-global-cryptocurrency-market-capitalisation-january-3-2018-1, accessed 11 October 2018.

71 Sue Walsh, "IBM, Maersk Roll Out Blockchain-Based Shipping Platform," *RTInsights*, https://www.rtinsights.com/ibm-maersk-roll-out-blockchain-based-shipping-platform, accessed 06 September 2018.

72 Kyunghee Park, "Blockchain Is about to Revolutionise the Shipping Industry," *The Economic Times*, https://economictimes.indiatimes.com/industry/transportation/shipping, accessed 20 April 2018.

73 Jorg Polzer, "Blockchain Technology: A Game Changer in Shipbuilding Industry," https://www.linkedin.com/pulse/blockchain-technology-game-changer-shipbuilding-industry-j%C3%B6rg-polzer, accessed 26 January 2018.

74 Jimmy Drennan, "Harnessing Tech Innovation from Blockchain to Kill Chain," *Center for International Maritime Security*, http://cimsec.org/harnessing-tech-innovation-from-blockchain-to-kill-chain/36996, accessed 11 July 2018.

75 "Global Claims Review - Liability in Focus," *Loss trends and emerging risks for businesses*, https://www.agcs.allianz.com/assets/PDFs/Reports/AGCS-Global-Claims-Review-2017.pdf, accessed 14 October 2018.

76 Vijay Sakhuja, "Autonomous Ship Operations in Need of New Regulatory Regime," *Society for the Study of Peace and Conflict*, https://www.sspconline.org/index.php/opinion-analysis/autonomous-ship-operations-need-new-regulatory-regime-thu-08302018, accessed 30 August 2018.

77 Jon Walker, "Autonomous Ships Timeline - Comparing Rolls-Royce, Kongsberg, Yara and More," *TechEmergence*, https://www.techemergence.com/autonomous-ships-timeline/, accessed 29 May 2018.

78 Hans Anton Tvete, "The ReVolt - A New Inspirational Ship Concept," *DNV GL*, https://www.dnvgl.com/technology-innovation/revolt/index.html, accessed 08 August 2016.

79 "ACTUV 'Sea Hunter' Prototype Transitions to Office of Naval Research for Further Development," https://www.darpa.mil/news-events/2018-01-30a, accessed 06 September 2018.

80 "Navy to Accelerate LDUUV and XLUUV Acquisitions," *Ocean News and Technology*, https://www.oceannews.com/news/defense/navy-to-accelerate-lduuv-and-xluuv-acquisitions, accessed 20 July 2018.

81 Stephen Chen, "China Developing Robotic Subs to Launch New Era of Sea Power," *South China Morning Post*, https://www.scmp.com/news/china/society/article/2156361/china-developing-unmanned-ai-submarines-launch-new-era-sea-power, accessed 22 July 2018.

82 "Shuttleworth Design: Mayflower Autonomous Research Ship," http://www.shuttleworthdesign.com/gallery.php?boat=MARS, accessed 08 August 2016.

83 Zhaki Abdullah, "Driverless Shuttle Trials Start on Sentosa; on-Demand Service for Public from 2019," *The Straits Times*, https://www.straitstimes.com/singapore/transport/driverless-shuttle-trials-to-start-on-sentosa-june-5-on-demand-for-public-from, accessed 25 September 2018.

84 Concerned with the behaviour of atomic particles.

85 They can simultaneously be in both states — '0' and '1'.

86 Entangled qubits behave together as a system — both take the same state.

87 They have a phase and can interfere with each other.

88 JM Gambetta, JM Chow, and Matthias Steffen, "Building Logical Qubits in a Superconducting Quantum Computing System," *Npj Quantum Information*, vol. 3, no. 1, https://doi.org/10.1038/s41534-016-0004-0, accessed 13 January 2017.

89 Ability to outperform a classical supercomputer on a defined computer problem.

90 Julian Kelly, "A Preview of Bristlecone, Google's New Quantum Processor," *Google AI*, http://ai.googleblog.com/2018/03/a-preview-of-bristlecone-googles-new.html, accessed 05 March 2018.

91 Zhang Zhihao, "Finding Paves Way for Even Better Computers," *Chinadaily*, https://www.chinadaily.com.cn/a/201808/18/WS5b77659da310add14f386736.html, accessed 18 August 2018.

92 "What Is Quantum Computing?", //www.research.ibm.com/ibm-q/learn/what-is-quantum-computing, accessed 29 October 2018.

93 Greg Satell, "The Industrial Era Ended, and So Will the Digital Era," *Harvard Business Review*, https://hbr.org/2018/07/the-industrial-era-ended-and-so-will-the-digital-era, accessed 11 July 2018.

94 Jacek Walendowski, Henning Kroll, and Esther Schnabl, "Industry 4.0, Advanced Materials (Nanotechnology)," *Thematic Paper*, https://ec.europa.eu/growth/tools-databases/regional-innovation-monitor/sites/default/files/report/RIM%20Plus_Industry% 204.0%2C% 20Advanced%20Materials%20%28Nanotechnology%29_Thematic%20paper.pdf, accessed 15 October 2018.

95 Schwab, *The Fourth Industrial Revolution,* ch.1, 17.

96 "The Challenges of Using New Materials in Shipbuilding," *Marine Offshore Technology*, http://www.marineoffshoretechnology.net/features-news/challenges-using-new-materials-shipbuilding, accessed 14 October 2018.

97 Bryan Nelson, "10 Ways Graphene Could Change the World," *MNN - Mother Nature Network*, https://www.mnn.com/green-tech/research-innovations/stories/10-ways-graphene-could-change-the-world, accessed 16 October 2018.

98 Warren Duffie Jr., "Navy Developing Ship Coatings to Reduce Fuel, Energy Costs - Office of Naval Research," https://www.onr.navy.mil/en/Media-Center/Press-Releases/2018/Navy-Developing-Ship-Coatings, accessed 21 June 2018.

99 Michael Grieves, "Digital Twin: Manufacturing Excellence through Virtual Factory Replication," http://innovate.fit.edu/plm/documents/doc_mgr/912/1411.0_Digital_Twin_ White_Paper_Dr_Grieves.pdf, accessed 25 October 2018.

100 Jake Frith, "'Digital Twins' Approach Could Cut Costs in Shipbuilding," *Maritime Journal*, https://www.maritimejournal.com/news101/vessel-build-and-maintenance/ship-and-boatbuilding/digital-twins-approach-could-cut-costs-in-shipbuilding, accessed 06 November 2018.

101 Gareth Evans, "The Digital Naval Shipyard," *Naval Technology*, https://www.naval-technology.com/features/digital-naval-shipyard/, accessed 12 February 2018.

102 Roland Freist, "General Electric Converts Diesel Ships into Smart Navy - Digital Twin," http://www.hannovermesse.de/en/news/general-electric-converts-diesel-ships-into-smart-navy-71809.xhtml, accessed 26 February 2018.

103 JE Siegel, S Kumar and SE Sarma, "The Future Internet of Things: Secure, Efficient, and Model-Based," *IEEE Internet of Things Journal* vol. 5, no. 4 (August 2018), 2386–98.

104 Computing is made to appear anytime anywhere using any device at any location.

105 E Sisinni et al, "Industrial Internet of Things: Challenges, Opportunities, and Directions," *IEEE Transactions on Industrial Informatics* vol. 14, no. 11 (2018), 4724-34.

106 Guy Daniels, "Inmarsat Introduces Smart Shipbuilding as Telcos Look to IoT Service Opportunities," *TelecomTV*, https://www.telecomtv.com/content/iot/inmarsat-introduces-smart-shipbuilding-as-telcos-look-to-iot-service-opportunities-15924/, accessed 05 September 2017.

107 "Internet of Everything - Capabilities for the US Navy," *White Paper*, https://www.cisco.com/c/dam/en_us/solutions/industries/us_government/resources/navy-ioe-wp1c.pdf, 27 October 2018.

108 Stuart Corner, "NZ Navy First with IoT-Equipped Warship," *PC World*, https://www.pcworld.idg.com.au/article/615707/nz-navy-first-iot-equipped-warship/, accessed 27 October 2018.

109 Stefano Zanero, "Cyber-Physical Systems," *IEEE Computer Society* vol. 50, no. 4 (April 2017), 14–16.

110 RH Rawung and AG Putrada, "Cyber Physical System: Paper Survey," *International Conference on ICT For Smart Society*, (2014), 273–78.

111 "DNV GL Technology Outlook 2025 – Shipping and Digitalization," https://to2025.dnvgl.com/shipping/digitalization/, accessed 27 October 2018.

112 A delivery method for mixed loads from different suppliers.

113 A lean method to manage and improve workflow which allows evolutionary changes whilst the work is in progress.

114 Karl Hribernik, "Industry 4.0 in the Maritime Sector," *Bremer Institute for Production and Logistics*, http://www.mlit.go.jp/common/001127983.pdf, accessed 27 October 2018.

115 Rahul Jain, "This Will Be India's Decade, Stars Are Well-Aligned," *The Economic Times*, July 3, 2023, Vol. 51 No. 130 edition, sec. ET Markets.

116 Dhiraj Nayyar, "American Graffiti on Indian Walls," *The Economic Times*, July 3, 2023, Vol. 51 No.130 edition, sec. The Edit Page.

117 PH Diamandis and Steven Kotler, *Bold: How to Go Big, Create Wealth and Impact the World*, (New York: Simon & Schuster, 2015), 182.

118 "Processing Power Compared: Visualizing a 1 Trillion-Fold Increase in Computing Performance," *Experts Exchange*, https://pages.experts-exchange.com/processing-power-compared, accessed 01 November 2018.

119 Leslie Feldman, "Cyber Security and the Fourth Industrial Revolution," *Symantec Corporation*, https://www.symantec.com/blogs/expert-perspectives/cyber-security-and-fourth-industrial-revolution, 28 April 2018.

120 Sakhuja, *Autonomous Ship Operations*, ch. 4.

5

Case Study – Japan

Introduction

Japan is one of the earliest countries in Asia to record significant economic growth based on export-oriented industrialisation after World War II. It is an archipelago of 4000 islands of which only four are sizeable ones, often referred to as the 'home islands'. Japan has a geographical extent of 1,200 miles; an area of 145,000 square miles; 17,000 miles of coastline and a population of 128 million, making it the tenth most populous country in the world. With virtually non-existent natural resources, it is the most affluent and economically productive nation after the US. Some of the defining characteristics of Japanese society include an agricultural base, prevalence of group loyalty, and well-established imperialism. Japan has developed its economy by importing necessary resources – including food – while exporting industrial products through its maritime transport system. This system carries about 96 per cent of supplies that enter and leave the country.

Japanese maritime business has three principal groups – shipping companies (operators and charterers), shipbuilders, and shippers (manufacturers). The private owners supply ships to international operators and charterers who have maintained long-term contracts for the carriage of cargoes with stable freight rates, with less speculative risk and long-term stability. The reason behind this successful characteristic maritime business model is the traditional Japanese trading system *Keiretsu* (group companies) and *Zaibatsu* (financial combines), which enable shipping companies to have reliable relationships with shippers and shipbuilders based on financial support provided by the banking system. Therefore, when shippers establish manufacturing plants in foreign countries, the shipping companies provide them with the necessary transport services.[1] This system has been at the root of a successful modern Japanese industrial complex.

History Before the Mid-Nineteenth Century

Japan has traditionally been a group of self-contained and semi-isolated islands off the Asian mainland. Historically, it borrowed cultural ideas from the Chinese civilisation. Japanese society was deeply influenced by external factors, including Buddhism, the Chinese script, and the Chinese system of centralised political and administrative patterns. Japan was never occupied throughout its history and was never invaded before 1945. Long before Western presence in these islands, the country's maritime trade extended to East and South Asian coastal waters. The population of Japan increased four-fold from 30 million in the 1850s to 120 million in the 1990s; it has now levelled off, with Japan facing the ageing population issue.

Japanese history is conventionally divided into six political eras. The first era consists of early indigenous culture from pre-historic times to 400 BCE. The second era consists of the three centuries of continental influence, primarily Chinese. The third era spanned from 700 AD to 1200 AD – an era of the synthesis of foreign and native ideas. The fourth era lasted from 1200 AD to 1600 AD; a period of military feudalism, characterised by the growth of militant groups, political factions, and civil wars. The fifth era was from 1600 to 1850 – the Tokugawa shogunate – mainly under three successive outstanding military leaders. This was the period of the political reunification of Japan. Lastly, the modern phase commenced in the mid-1850s, marked by the arrival of Commodore William Perry.[2]

The two-and-a-half centuries of the Tokugawa Period (1603-1857) was an era of stability, seclusion, and conformity. Around 1636, under Shogunate rule, Japan closed its doors to the world. Japanese citizens were not allowed to travel abroad, and the construction of large ocean-going vessels was prohibited. It was only after about two-and-a-half centuries of this period that Japan opened up to the world, following the transfer of power from the shogun Tokugawa Yoshinobu in 1867 when it effectively became an imperial state.[3] The most significant impact of this period was that it transformed the adventurous and spirited people of earlier centuries into a regimented nation by the nineteenth century. The society became structured and rigid; however, the outdated feudal outlook continued to be preserved. It is widely believed that the attitudes formed during this period were one of the contributing factors towards Japan accepting Western norms more easily than any country in Asia; after opening up to Western powers, it forged ahead of its continental neighbours. Paradoxically, the Tokugawa legacy proved to be both a boon and a bane in the following years.[4]

In 1720, the eighth Tokugawa shogun Yoshimune (1716-1745) lifted the ban on teaching Western subjects in schools, and introduced courses in Shipping, Foreign Affairs, and Conservation of Natural Resources. These initiatives acted as early seeds of transformation that eventually led to the opening up of Japan. Around 1798, some Samurais began plans to modernise using Western technologies in gunpowder, metals, and ships. They initiated ideas of seizing land around the Japanese perimeter, including Korea, Manchuria, and eastern Siberia. Furthermore, they advocated for the establishment of a unified national army command, a strong national government, the end of feudalism, and a robust economic base. As early as 1809, Sato Nobuhiro (1769-1850) argued in 'Maritime History of Nations' that the greatness of nations lay in commerce and the navy.[5]

Mid-Nineteenth Century to the End of World War II

In 1868, Japan came under the direct imperial rule of Emperor Meiji, marking an end to the military rule of the Tokugawa Shogunate and the commencement of the modern phase of Japan. The transformation, however, had been triggered by the arrival of four US Navy ships under Commodore William Perry in Edo Bay – later named Tokyo – on 08 July 1853. He subsequently returned in February 1854 with eight ships to sign the Kanagawa Treaty on 31 March 1854. The treaty facilitated the opening of the Japanese market to the Americans, leading to greater interactions.

In the years following the signing of the Kanagawa Treaty, the British, French, Russians, and the Dutch also signed similar agreements. The Meiji Restoration period has often been classified into two phases. In the first phase from 1868 to 1890, Japan focussed on internal consolidation to streamline governance, financial systems, and industries through the *Zaibatsu* (cartels).

The Emperor bestowed the first constitution on 11 February 1889, installing a representative government. The country adopted universal education that transformed Japan into the first Asian country with a literate populace, leading to its concurrent rise as a major industrial power with formidable military strength. Japan pioneered authoritarian techniques of using education as a means of political tutelage. Meiji leaders believed that Japan could only resist Western powers' domination by building military power. The Japanese army modelled itself after Prussian organisation and discipline, while the navy established training stations with English advisers by 1869.[6]

In the second phase, Japan intensified industrialisation to achieve higher growth in sectors like chemicals, fertilisers, cotton weaving mills, electrical industries, coal and steel factories, and shipbuilding. Among these, the shipbuilding industry was central to the opening up and expansion plans from the outset. This was because, during the initial interactions with Commodore Perry, the ruling shogunate realised American ships were of significantly higher quality compared to locally-built Japanese ships, highlighting Japan's relative technological backwardness. However, the initial phases consisted of outright procurement of ships built overseas.

Japan set out to build up its military power in 1877. Since the indigenous shipbuilding industry was nearly non-existent at the time, Japan approached foreign shipyards and placed orders with British shipyards for several armoured ships, which were delivered in the 1880s. In 1882, the First Naval Expansion Bill was passed, which provided for the construction of 48 warships over a period of eight years. The Bill also allocated funds to develop the domestic naval industry. Several government shipyards, including those at Yokosuka, Kure, Maizuru, and Sasebo, were established, while private shipyards like Onohama took on some of the orders. In addition to the shipbuilding industry, the domestic steel industry was also developed.

The capacity of shipbuilding was further enhanced under the ten-year Naval Expansion Bill of 1896, and the Japanese shipbuilding industry continued in the growth trajectory as the nation continued its pursuit towards a blue-water navy. The government also introduced subsidies for shipping companies to operate on new routes to Europe, North America, Australia, and India. The subsidy was fixed at a lower rate for foreign-built vessels than for those that were constructed in Japan. This decision had a significant impact on encouraging the domestic shipbuilding industry, which was further reinforced in 1899 when foreign-built ships were prohibited from participating in coastal trade. Specific measures to support the Japanese shipbuilding industry effectively date from this time.[7]

During World War I, there was a huge demand by the Allies for the Japanese to supply warships and to participate in the massive merchant shipbuilding programme. Many new shipyards were established, while the existing shipyards were expanded to meet this upsurge in the requirement to export warships and merchant ships. By 1919, Japan had built several hundred ships for France and Great Britain, and its own navy and merchant marine had grown significantly. The Japanese Navy was no longer dependent on foreign sources for either its warships or their components.[8] It was at this stage that Japanese shipbuilding reached international standards and became competitive.[9]

In 1920, the Imperial Japanese Navy (IJN) was the third largest navy after Great Britain and the US. They acquired warships from Britain and France and also built them with the support of the French. Japan had four major naval arsenals at Yokosuka, Kure, Sasebo, and Maizuru, where its naval shipbuilding yards were located; these were owned and operated by the IJN. Many private shipbuilding companies were also well established by this time. Ishikawajima, which was established in 1853, later merged with Harima Docks and is popularly known as Ishikawajima-Harima Industries (IHI). In 2013, it merged with the Universal Shipping Corporation (USC) to form the Japan Marine United Corporation (JMU).[10] Kawasaki established private yards in Tokyo and Kobe, in 1878 and 1886, respectively. The Kawasaki yards built numerous warships such as destroyers, aircraft carriers, and submarines for the IJN. The Nagasaki shipyard was established along with a foundry in 1857. This yard was leased to Mitsubishi in 1884 and is now a part of Mitsubishi Heavy Industries.

In the pre-World War II period, the Japanese shipbuilding industry had established itself as a source of product innovations in ship design. In 1916, Japan commissioned the battleship *Nagato*, which introduced the sixteen-inch gun to the navies of the world. In 1918, Japan launched the so-called "8-8" programme, which envisioned the construction of eight battleships and eight battlecruisers by 1928. The Washington Naval Treaty of 1922 – also known as the Five-Power Treaty[11] – posed restrictions by limiting the sizes and capabilities of warships that were permitted to be built by Japan. In 1934, Japan served notice to withdraw from the Washington Treaty and continued massively expanding its shipbuilding capabilities, only limited by a shortfall in resources because of American embargoes. However, between 1934 and the start of the Second World War, Japanese shipyards had constructed eighty-three warships, with 42 per cent of them being built in government yards.[12]

After World War II

At the end of World War II, Japan surrendered and came under the administrative control of the Supreme Commander of Allied Power (SCAP) – led by the US, with the inclusion of representatives of all major allies. Initially, SCAP aimed to demilitarise and democratise Japan. Accordingly, SCAP decentralised governance, streamlined the election of members to both houses of the Diet and implemented a universal franchise. It reformed the education system to align it with the US system of schooling, colleges, and universities. It also implemented extensive land reforms to abolish the feudal system. In 1948, the US changed the mandate of SCAP from reform to recovery. When the Korean War broke out in June 1950, the US expedited the recovery process of Japan and changed its status to an ally of the US, with full sovereignty, in a treaty signed in San Francisco in September 1951. As a part of the treaty, the US offered to loan the Japan Maritime Self-Defence Force (JMSDF) eighteen frigates and fifty Landing Ship Logistics (LSLs), the first of which was transferred in January 1953. The focus then shifted towards evaluating the domestic shipbuilding capabilities to provide for future requirements of warships for the JMSDF.

During the war, although Japan suffered massive damages to its ships and submarines – including its merchant fleet, 85 per cent of its shipbuilding capacity remained undamaged. The surviving facilities included 50 shipyards, mostly in the private sector which possessed 126 berths, and 75 drydocks, and were organised by 35 firms.[13] However, little progress could be made due to shortages of material, power, and skilled workforce. Once the US' stance on Japan changed, SCAP approved a programmed shipbuilding scheme and new constructions began in the Japanese shipyards.[14]

From 1945 to the 1960s, Japan maintained a low profile in international affairs. During this period, it received massive US aid which was further increased during the Korean War. Japan used the US aid to construct new industrial plants. There was a strong mutual aid bond between the government and industry. There was also a pronounced work ethic, a high level of technology, and political stability. All these factors helped Japan register a 9% growth during the period between 1950 and 1955. Japan sustained this momentum further and used its self-generated affluence to achieve a growth of 12% from 1955 to 1960. The growth story of Japan continued, and it achieved growth rates of 10% in the golden 1960s, 5% in the 1970s, and 4% in the 1980s. However, the Japanese economy only grew at a rate of 1.7% in the 1990s – also referred to as the 'lost decade', and it maintained this rate until the global economic crisis in 2008-2009.

In the period after the Second World War, a significant development in global shipbuilding impacted the Japanese shipbuilding industry. There was a slump in demand for American-built ships and many shipyards in the US that were set up during wartime were being closed. Amongst these were seven Kaiser emergency shipyards owned by Henry J Kaiser, which were the frontrunners in introducing innovative production technologies like Group Technology (GT),[15] pre-outfitted sub-assemblies, etc.

When Henry Kaiser decided to exit the industry, Elmer L Hann – who had previously served as the General Superintendent at his Swann Island yard in Oregon – relocated to Welding Shipyard in Norfolk, Virginia. He also assumed the role of the principal advisor on ship construction to National Bulk Carriers Inc (NBC) of New York. NBC wanted to build large iron-ore vessels for Venezuelan trade and tasked Hann to survey potential sites. Although the search for a suitable facility in Germany and the UK proved unsuccessful, in 1950, Hann learned that a portion of the Japanese Naval Shipyard at Kure could be made available if he found it suitable. The yard, with a 150,000 DWT building dock and a hundred-ton crane, was deemed suitable.

In 1946, SCAP authorised the Japanese Shipbuilding Bureau to lease the Kure shipyard to the NBC for ten years, at a nominal rent. The NBC agreed to use Japanese steel and allow Japanese engineers to access the facility to absorb technology. The NBC appointed Hisashi Shinto as Chief Engineer at Kure. Shinto applied all the technical knowledge gained by the Americans during the war and further improved production by implementing ideas, including those of automation and planning of assembly line production used in the aircraft industry. Kure Shipyard emerged as the world's most technically advanced facility of its kind, serving as a reference for shipyards throughout Japan. It produced high-quality vessels at competitive prices for eleven years. Approximately 4000-5000 engineers visited the production facility and absorbed the technology. As a result, output across Japanese shipyards expanded and Japan became the world's largest producer of merchant ships by 1956.[16]

The transfer of technology was within the ambit of a broader alliance with the US. It was in line with the Yoshida doctrine, formulated by Shigeru Yoshida – the Prime Minister of Japan between 1946 and 1954. The doctrine emphasised Japan's economic growth as the primary objective, with no involvement in international strategic political issues and the provision of military bases to the US.[17] The alliance provided technology as well as economic assistance to help Japan recover its economy. The lease of Kure yard by NBC helped Japanese shipbuilders learn to organise their work in accordance with GT's principles, adopt advanced welding techniques with minimum distortion to the structure, and appreciate the advantages of having educated middle-level managers who are trained in the complete shipbuilding system.

In the 1950s, Dr W Edwards Deming introduced Statistical Control Methods (SCM) in the Japanese industry, which radically improved the quality of finished products. The application of SCM in shipbuilding provided meaningful indicators of how work is to be executed. For the first time, it was possible to evaluate the impact of innovation on the work process. This gave rise to the Quality Circle and as a result of this, employees at all levels in Japanese shipyards started participating in quality improvement on a daily basis.

Dr Hisashi Shinto, who is known as the founding father of modern Japanese shipbuilding, has contributed immensely to the Japanese shipbuilding industry. He initially worked with Elmer Hann as his Chief Engineer. After Hann's departure from Japan, Shinto became the head of the IHI shipyard at Kure. Shinto, who was impressed with the production control procedure used in the aircraft manufacturing industry, developed an improved shipbuilding system based on it, incorporating the best practices of Kaiser's techniques and SCM. He developed an improved shipbuilding system, known as the flow-line method – which is a combination of block construction using welding, along with the application of project scheduling. This enabled a worker to achieve in one man-hour the equivalent of three man-hours of work in a traditional US shipyard.

The IHI Kure yard became the first Japanese shipyard to adopt the flow-line method. This method brought in higher outputs and faster production, mainly because of the scheduling of work which helped in producing blocks at the right time, in the right order, and in the right quantity – this facilitated the production line to flow smoothly. The IHI Kure yard could routinely build a vessel from keel-laying to launching in five months with the flow-line method. This method of building large vessels at a fast pace was absorbed by other Japanese shipyards very soon. Mitsubishi's Nagasaki Shipyard received the block construction technique from Kure, and they reorganised themselves around the flow-line production method in 1956, which helped them achieve better results.

Another improvement in Japanese shipyards compared to the American system of shipbuilding was the adoption of zone outfitting instead of pre-outfitting. In pre-outfitting, the equipment and system are outfitted system-wise, whereas, in zone outfitting, all equipment and systems pertaining to a location/zone are outfitted together in the block. A method was evolved to undertake painting of the ship in zones, which was known as the 'zone painting method'; this improved the surface preparation and quality of paint coating. All these approaches helped the yard accomplish a greater degree of work on the shop floor in a cleaner and more comfortable working environment. Adoption of the zone outfitting technique resulted in the grouping of the system/equipment independent of their function. This was a result of GT, which was used in mass production. GT also led to the formulation of the Product Work Breakdown Structure (PWBS), a system of grouping products by similarities in production, without regard for the end use of the system. With time, the PWBS has evolved and today it is used for time and cost estimation in most of the new shipbuilding projects.

In Japanese shipyards, the 'planning phase' predominates. The planning department works closely with the design, material identification, and procurement departments for every new project executed by them. Procurement orders for raw materials and equipment are placed while the design is in progress; specifically, orders for 70 per cent of the material are placed when 30 per cent of the ship design is complete. As a result, the material reaches the yard before production starts. This helps shipyards drastically reduce the time between signing the contract and delivering the ship.

The transfer of modern shipbuilding technologies not only expanded the production capacity of the Japanese shipbuilding industry but also enabled the entire industry to be reorganised so that it was able to respond to changes in demand as they occurred. When the closure of the Suez Canal led to increased global demand for larger vessels that could handle longer routes around the Cape of Good Hope by taking advantage of economies of scale, the Japanese shipbuilders could respond quickly and take the lead in the market. Later, when a slump followed, Japan could employ some of its excess capacity towards carrying its growing overseas trade. Between 1955 and 1969, Japan's exports increased from US $2 billion to US $6.6 billion, and imports from US $2.4 billion to US $7.9 billion. Since much of the imports consisted of bulk cargoes such as iron ore and crude oil, the types of vessels used for these purposes made up a significant portion of the Japanese fleet.

During the 'oil shock' in the early 1970s, there was a significant drop in ship demand, severely impacting the Japanese shipbuilding industry as Japan was the world's largest ship producer. The industry resorted to restructuring so that it could reduce the cost and time of construction. The 17 original firms that had emerged in the post-war era were merged into seven new groups listed in Table 5.1. As demand continued to decrease, shipbuilding capacity was further reduced by 33 per cent by 1978-1979. The closure of shipyards was smooth since the workers were shifted to other businesses within the *Zaibatsu*, and a major chunk were daily wage labourers who were utilised by the sub-contractors in other works.[18]

Table 5.1: **Major Shipbuilding Groups in Japan - Established in 1971**[19]

Ser	Major Shipbuilding groups established in 1971
1	Mitsubishi Jyukogyo (Mitsubishi Heavy Industries)
2	Mitsui Zosen (Mitsui Shipbuilding)
3	Sumitomo Jykikai Kogyo (Sumitomo Heavy Machinery)
4	Kawasaki Jyukogyo (Kawasaki Heavy Industries)
5	Hitachi Zosen (Hitachi Shipbuilding)
6	Ishikawajima Harima Jyukogyo (Ishikawajima Harima Heavy Industries)
7	Nippon Kokan (Japan Steel Pipe)

During the global oil energy crisis in the 1970s, the multi-polar world order was undergoing a re-arrangement. Japan played an active role, emphasising the economics of diplomacy. In order to sustain its industrial and commercial expansion, Japan focussed on increasing exports, particularly in sectors like heavy and chemical industry. Japan focussed on US markets as well as Asian markets and could trade with all blocks – from China to Taiwan. A third of its total labour force was industrial. The country had inherited a traditional entrepreneurial spirit dating back to the Tokugawa era when the first Japanese companies were established. Over the centuries, these companies grew to become behemoths in the twenty-first century. The spectrum was wide and included chemicals, consumer electronics, iron and steel, copper, petrochemicals, pharmaceuticals, shipbuilding, aerospace, textiles, and processed foods. The per capita zoomed from $500 in the post-war period to $33,000 in 2006.

The Warship-Building Industry After World War II

By the early 1950s, the Japanese shipbuilding industry had grown in capability to participate in the construction programmes for the Japan Maritime Self-Defence Force (JMSDF) and the Maritime Safety Agency (now called Japan Coast Guard – JCG). However, due to restrictions imposed by the Japanese constitution prohibiting the building of military capabilities and also imposing financial limits to government spending on these sectors, there was little progress in warship construction. The situation changed after the American Treaty of 1951 when the US offered to loan Japan eighteen frigates and fifty LSLs, the first of which was transferred in January 1953. At the end of 1954, Japan resumed the construction of its own warships, laying down two destroyers and two frigates. Since no government dockyards had survived the war, these ships and those that followed were constructed in private shipyards. The Japanese merchant shipbuilding industry – which by this time was a global leader – took on the warship and submarine orders required, for both the defence capability build-up and expansion plans of the JMSDF and the JCG. Those yards that undertook the construction of the JMSDF's vessels had already accumulated invaluable experience in modern processes and technologies by building merchant ships.

Since the mid-1950s, Japanese shipyards have evolved from constructing destroyers and frigates to building and supporting a naval force with both blue-water capability and a force structure that includes destroyers, frigates, and submarines. Moreover, the JMSDF now deploys more major surface combatants than the Royal Navy (RN) and the French Navy. The Japanese government has kept the naval shipbuilding industry busy with a steady stream of orders for major vessels like destroyers, frigates, and submarines, as well as numerous patrol, mine warfare, amphibious, and auxiliary vessels for both the JMSDF and the JCG. Most of the naval contracts have been given to six major firms, namely MHI, KHI, IHI, MES, Hitachi Zosen, and SHI. The Japanese government has carefully allocated the orders to these firms, with the objective of keeping them all in the warship-building business.

Japan's military-industrial complex – particularly its warship-building capability – has greatly benefitted from its economic rise and advancement in technologies. Limitations, if any, are self-imposed, like the maximum defence expenditure limited to one per cent of national wealth (although the amount is still much larger than most other nations' expenditure). Further, there is an embargo on acquiring offensive capabilities like cruisers, nuclear submarines, aircraft carriers, etc.; the JMSDF can have a maximum of 16 submarines at any given time, and defence exports are prohibited. The JMSDF has found ways to circumvent the proscription of major warships by ordering warships with enhanced capabilities.

In 1993, the JMSDF ordered the construction of amphibious ships – nominally a tank landing ship (LST), which would function as a helicopter carrier in other countries. Two *Hyuga*-class helicopter destroyers were built by IHI Marine United, Yokohama, Hyuga in 2009 and Ise in 2011. Two *Izumo*-class helicopter destroyers with VSTOL were built by the same Izumo Yard in 2015 and Kaga in 2017. Similarly, since cruisers are not permitted for induction, the JMSDF ordered large destroyers, including the 9485-ton *Kongo*-class guided missile destroyers – built around the Aegis-integrated weapons system of Lockheed Martin, inspired by USN's *Arleigh Burke* DDGs. Since the maximum number of submarines in active service is limited by budget constraints, the JMSDF retires its submarines at a much earlier age than is typical in other countries in order to replace them with newer submarines featuring contemporary technology. This is done to keep regular, spaced-out orders on shipyards so that the submarine-building capability is sustained.

Since there is a ban on weapons exports, these submarines are consigned to scrapyards. For example, KHI has built 28 submarines since World War II and recently commissioned the tenth *Soryu*-class submarine in March 2019. It is a diesel-electric attack submarine (SSK). The ninth *Soryu*-class SSK was commissioned in March 2018, built by MHI, and was the fifth *Soryu*-class to be built by MHI. All these SSKs have air-independent propulsion (AIP) systems and the next two will have lithium-ion batteries.[20]

Presently, the JMSDF has 154 ships and submarines, with weapons of the latest technology and propulsion which rides on the integrated merchant-shipbuilding capability of Japan. Technological developments in global shipbuilding and Japan's leadership position translate into a significant advantage to the JMSDF and JCG. In turn, they support the industry with regular orders to keep the funds flowing.

Recent Developments in Shipbuilding

As per some estimates, the global volume of ocean-borne cargo movement has increased by 50 per cent in the first decade of the twenty-first century, and the trend is expected to continue.[21] Considering that Japanese international shipping lines source 90 per cent of their ships from Japan, the increased maritime transport presents a significant opportunity for Japanese shipbuilders. In addition, the government of Japan is also investing in transforming Keihin and Hanshin ports into major hubs to utilise the growing maritime transport. The Ministry of Land, Infrastructure, Transport and Tourism (MLIT) has set a goal to reduce GHG emissions by 30 per cent from international shipping. This adds a fillip to growth in green technologies and LNG-fuelled ships, with Japan expecting international demand for these highly efficient vessels to grow further. Japanese-listed companies have doubled their investments in R&D from 2006 to 2011, reaching US $157 million.[22]

The Japanese government introduced a series of measures – such as bond purchases and increased government spending – to revive the economy, leading to a significant depreciation of the yen against the US dollar in 2012-13. Since most shipbuilding is priced in US dollars, the local currency's depreciation improved pricing for buyers, making Japan's costs comparable to those of South Korea and China, with Japanese reliability and consistency being considered superior. This development is significant since it was happening at a time when overcapacity was driving reorganisation and consolidation in China. Japanese shipbuilding is strategically transitioning from the former mass production of long-series standard bulk carriers and oil tankers to short-series, technically advanced high-end units. Some analysts view this shift as a renaissance in Japanese shipbuilding.[23]

Some of the R&D efforts are jointly undertaken by companies; for example, ClassNK is working with Imabari and Sanoyas Shipbuilding Corporation to test technologies that meet new EEDI[24] regulations in the actual operating environment. In September 2014, Mitsui Engineering and Shipbuilding Co. commenced work on a bulk carrier, which will be its first large eco-friendly ship. The vessel will be 25 per cent more fuel-efficient than conventional bulk carriers. KHI is developing Liquid Hydrogen carriers which will be the first-of-its-kind to ship liquid hydrogen – extremely volatile cargo. Liquid hydrogen is produced in Australia by the gasification of brown coal and shipped to Japan, to fuel the next generation of hydrogen-fuelled carbon-free automobiles. Toyota and Honda have been developing hydrogen-fuelled cars, while the Japanese government has invested US $100 million to convert existing taxis into hydrogen-fuelled vehicles to showcase the technology to the world during the 2020 Tokyo Olympics.[25]

Japanese shipbuilders are focussing their R&D efforts on energy-saving and environmental technologies, with a focus on reducing CO_2 emissions and improving fuel efficiency. MLIT envisions that the innovations in green technologies can be utilised by other industries as well. To address competition, the Japanese maritime industry has also diversified its products across various shipyards. For example, MHI's Kobe Shipyard ceased to construct merchant vessels and is now constructing offshore units, research vessels, and cruise ships. Some other shipyards are focussing more on shipbuilding services, including providing design, technology support, or training engineers.[26]

As per IHS Markit's new building statistics for 2017, Japan is in the third position after China and South Korea, with a market share of 26.1 per cent. Japan has made plans to reinvigorate its shipbuilding industry and increase its market share to 30 per cent by 2025. In 2016, MLIT defined four driving forces to achieve Japan's market share goal. First, yards must develop sophisticated products and services, using big data, automation, and the Internet of Things (IoT), as well as sophisticated ship design methods to develop new vessel types. Second, yards must cultivate new businesses, like offshore businesses, and transport liquified hydrogen to develop new skills in the workforce. Third, they must acquire advanced manufacturing processes that will enable them to visualise and share each element of the vessel construction process with partners. Fourth, they must develop HR to take advantage of new technologies.[27]

In early 2018, Japanese shipyards produced about half the world's LNG ships. KHI's Sakalde Yard completed the construction of the largest Moss spherical tank LNG ship as well as a dual-feed diesel-electric propulsion (DFDE). In March 2018, another Moss spherical ship was delivered by the Nagasaki Yard of Mitsubishi Heavy Industries (MHI). Imabari has also achieved considerable success in LNG as well as next-generation ultra-large container ships.[28]

Nippon Yusen has been working with Norwegian company DNV GL on technologies that enable ships to use data to assess collision risk. In fact, Nippon Yusen, along with Mitsui OSK Lines, is working with Japan Marine United and other shipbuilders to build self-piloting cargo ships. These ships will be smart ships, using Artificial Intelligence to plot the safest, shortest, and most fuel-efficient routes, and are expected to be in service by 2025.[29]

During the period of rapid growth in the world economy – from the 1950s through the 1980s, Japan was the powerhouse in the manufacturing industry, introducing revolutionary production improvement systems – including Kanban[30] and the Toyota Production System (TPS).[31] Therefore, Japan was expected to be the natural leader in the onset of Industry 4.0. However, the leading countries in implementing the concept of Industry 4.0 have been Germany and the US. The Japanese initiative towards the transition to Industry 4.0 is the Industrial Valuechain Initiative (IVI), which is a forum to design a new society by combining manufacturing and information technologies to enable all enterprises to progress in collaboration. The IVI aims to facilitate a transition from human-centric manufacturing to a mutually-connected system architecture using IoT. The IVI formulates a loosely defined connectivity model that enhances the value of each enterprise as a cyber-physical production system.[32] Almost all major shipyards are part of this initiative.

On 24 July 2019, in a ceremony marking one year until the start of the summer games in Tokyo, officials promised a high-tech and eco-friendly Olympics event. The Olympic organisers unveiled medals made from recycled materials collected from old electronics.[33] Japan is also working on displaying hydrogen-fuelled automobiles for ferrying athletes to the venue. The country's underlying focus is precisely the strategy of its shipyards to recapture their share of the global shipbuilding market, which includes the high technologies of Industry 4.0 and green technologies. Shipyards are ushering in the concepts of connected and collaborated manufacturing under the IVI. They are focussed towards building high-end ships like LNG and CNG carriers. There is also a drive to build liquid hydrogen carriers which will carry hydrogen from Australia to Japan to meet the requirements of hydrogen-fuelled automobiles. The manufacturing industry, with shipbuilding as a major component, powered Japan's rise in the global economic order. Today, its shipyards are focussing on advanced and green technologies to maintain their leadership position.

NOTES

1   Keiji Habara, "Maritime Policy in Japan," *Journal of Maritime Research,* vol. 1, no. 1 (March 2011), 65–84.

2   MW Meyer, *Japan: A Concise History* (United States: Rowman & Littlefield, 2012), 21–22.

3   Davies, *Japanese Shipping and Shipbuilding in the Twentieth Century,* ch.1, 43–44.

4   Meyer, *Japan: A Concise History,* ch.5, 180.

5   Meyer, *Japan: A Concise History,* 118.

6   Meyer, *Japan: A Concise History,* 150–56.

7   Davies, *Japanese Shipping and Shipbuilding in the Twentieth Century,* ch.1, 50.

8   Todd and Lindberg, *Navies and Shipbuilding Industries,* ch.1, 136.

9   Davies, *Japanese Shipping and Shipbuilding in the Twentieth Century,* ch.1, 44.

10 'Japan: IHI Marine United, Universal Shipbuilding Merger Creates Giant', https://world maritimenews.com/archives/114922/japan-ihi-marine-united-universal-shipbuilding-merger-creates-giant/, accessed 04 January 2013

11 The treaty was signed among the major nations that had won World War I, which agreed to prevent an arms race by limiting naval construction.

12 Todd and Lindberg, *Navies and Shipbuilding Industries,* ch.1, 137.

13 Tomohei Chida and PN Davies, 'The Japanese Shipping and Shipbuilding Industries', *International Journal of Maritime History* vol. 2, no. 2 (01 December 1990), 241–45.

14 Davies, *Japanese Shipping and Shipbuilding in the Twentieth Century,* ch.1, 86.

15 A manufacturing philosophy in which parts having similarities — geometry/manufacturing process and/or function — are grouped and processed together. This reduces the production time and improves quality.

16 Davies, *Japanese Shipping and Shipbuilding in the Twentieth Century,* ch.1, 106-108.

17 Christopher W Hughes, *Japan's Re-Emergence as a 'Normal' Military Power*, (New York: Routledge, 2006), 21.

18 Davies, *Japanese Shipping and Shipbuilding in the Twentieth Century*, ch.1, 88-95.

19 Davies, *Japanese Shipping and Shipbuilding in the Twentieth Century*, ch.1, 94.

20 Franz-Stefan Gady, 'Japan Commissions 10th Soryu-Class Diesel-Electric Attack Submarine', *The Diplomat*, https://thediplomat.com/2019/03/japan-commissions-10th-soryu-class-diesel-electric-attack-submarine/, accessed 27 March 2019.

21 Habara, "Maritime Policy in Japan", 65-84.

22 Council Working Party on Shipbuilding Report 2013.

23 Sumanta Panigrahi, 'Asian Shipbuilding: A Dynamic Market', *GTR Export Finance Supplement*, http://www.citigroup.com/transactionservices/home/trade_svcs/docs/asian_shipbuilding.pdf, accessed 15 July 2019.

24 Energy Efficiency Design Index is mandatory for new ships, mandated by the IMO in July 2011. It aims to promote more energy-efficient and less polluting engines.

25 'Kawasaki Ship Designs Support Japan's Hydrogen-Society Plans', *Riviera Maritime Media*, https://www.rivieramm.com/news-content-hub/news-content-hub/kawasaki-ship-designs-support-japans-hydrogen-society-plans-29850, accessed 17 July 2019.

26 '2015-Marine-Industries-Resource-Guide-Japan-and-China.Pdf', *Maine International Trade Center*, http://www.mitc.com/wp-content/uploads/2015/04/2015-Marine-Industries-Resource-Guide-Japan-and-China.pdf?3dc2e8, accessed 15 July 2019.

27 Nick Savvides, 'Outlook 2018: Asian Shipyards to Embrace Innovation in 2018', *IHS Markit Safety at Sea*, https://safetyatsea.net/news/2017/outlook-2018-asian-shipyards-to-embrace-innovation-in-2018/, accessed 07 January 2019.

28 'Japan Prominent in Latest LNG Fleet Surge', *Riviera Maritime Media*, https://www.rivieramm.com/opinion/opinion/japan-prominent-in-latest-lng-fleet-surge-24777, accessed 18 July 2019.

29 Dave Lee, 'Self-Navigating Cargo Ships "by 2025"', *BBC News*, 9 June 2017, sec. Technology, https://www.bbc.com/news/technology-40219682, accessed 18 July 2019.

30 A process that enables smooth communication between the different stages of the production process.

31 Maximising the flow of goods while simultaneously minimising storage levels.

32 Yasuyuki Nishioka, 'What's IVI? – Industrial Valuechain Initiative', https://iv-i.org/wp/en/about-us/whatsivi, accessed 22 July 2019.

33 'The Final Countdown', *The Economic Times*, 25 July 2019, sec. Sports: The Great Games, 16.

6

Case Study – South Korea

Introduction

South Korea is a classic 'rags-to-riches' case and an economic miracle in the post-World War II era. It is a country located in the southern half of the Korean peninsula, with an area of approximately 100,000 square kilometres and an extensive coastline of about 2,400 kilometres. Korea was one of the poorest countries, with an underdeveloped agrarian economy largely dependent on foreign aid in the 1950s. In three decades, it transformed into a developed economy in what some refer to as the 'miracle on the Han River'.[1] South Korea achieved this transformation using a model similar to the one Japan followed during the Meiji Restoration and following the Second World War. However, the pace at which South Korea achieved this transformation stands out.

During the initial phase of transformation, the Republic of Korea (ROK) implemented extensive land reforms, which included land redistribution in such a way that individual farmers came to own small plots in which they did 'family farming'[2] – this became the primary method of farming in the country. Similar land reforms were implemented in Japan during the Meiji Restoration, following which the overall agricultural yield increased substantially. The increased agricultural yield improved food self-sufficiency and enhanced the purchasing capacity of a large percentage of the population. In subsequent phases, the government steered industrial growth by facilitating trade and commerce in such a way that large business conglomerates – called *chaebols* – could establish themselves in the global market through exports. The ROK planned its growth trajectory wherein the government steered development through export-oriented manufacturing and executed it through the disciplined private sector in the generative sectors[3] of steel and shipbuilding.[4]

History Before the Mid-Nineteenth Century

Around the third century CE, a variety of tribal people are believed to have settled in the Korean peninsula. They were the earliest human inhabitants of the peninsula, eventually forming an indigenous literate state. In the subsequent period, three independent states – Silla, Paekche, and KoguryO – competed for supremacy in the peninsula during the 'Period of Three Kingdoms' (fourth century CE to 676). In the seventh century, most of the peninsula was unified under Silla's rule. Thereafter, the Silla dynasty ruled from 676 to 935 AD and was followed by the rule of the KoryO (935-1392) and ChosOn (1392-1910) dynasties.[5] Each of these dynasties lasted long because the hereditary families and lineage of the ruling class provided stability. Foreign powers – especially China, some of the invaders like the Khitan, Jurchen, and Mongols (from the tenth to the thirteenth century) and the Manchus in the mid-seventeenth century – greatly influenced Korean society.[6]

Korea was a very stable society under the ChosOn dynasty (1392-1910). This was because there existed an equilibrium between the monarchy and the aristocracy: the former governed through a centralised bureaucratic setup, and the latter maintained a hierarchical social system dominated by the 'yangban' elite.[7] From 1392 onwards, Korea maintained a tributary relationship with China's Ming dynasty. Since the Japanese invasions of the late sixteenth century and the Manchu invasions of the seventeenth century, Korea completely isolated itself from the rest of the world. There were severe restrictions on foreign contacts, and no one was permitted to travel abroad unless approved by the State. The isolation was so extreme that Korea was often referred to as the 'hermit kingdom'.[8]

In the early nineteenth century, there was a substantial decrease in agricultural output in Korea. This began to strain the equilibrium between the monarchy, aristocracy, and the peasants which had hitherto kept the society stable. The resentment amongst the peasants led to a series of popular uprisings from 1811 onwards for the rest of the century.

Mid-Nineteenth Century to the End of World War II

Major uprisings like the Tonghak (Eastern Learning) and peasants' rebellions,[9] towards the end of the nineteenth century, continued to destabilise Korea. After the Meiji Restoration, the Japanese began to exercise greater influence in the region and began to engage with the Chinese, Americans, French, and British. Japan began to trade with Korea in the 1870s and forced it to open its ports and use Japanese currency. The developments veered to conflicts and eventually, Japan became the dominant power, initially pushing Kabo reforms[10] in 1894, and subsequently colonising Taiwan and Korea.[11] In 1905, Japan declared Korea as a protectorate and finally annexed it in 1910.[12]

After its annexation in 1910, Korea came under Japanese colonial rule till 1945. Although this period was one of oppression and Korean society had to undergo profound subjugation, it was also a period where the foundation was laid for Korea's future development. The Japanese oversaw the setting up of modern infrastructure befitting an industrial society, the establishment of a high standard of government efficiency, and the introduction of a modern school system of education. The government carried out extensive land surveys, and identified the extent of land for agriculture and its registered ownership by individuals. Those plots for which clear ownership could not be established were brought under government control. This was done to streamline land tax. The Japanese, along with the Americans, commenced the construction of railway lines in 1896, extending rail connectivity to virtually all the ports and cities by 1945. The government introduced commercial agriculture, primarily to meet Japanese demands for rice, soybean, silk, and sheep.

A network of mines of gold, iron, and steel, in addition to many industries, were set up to feed through the developed logistics chain formed by the network of railroads feeding the ports. In 1931, Japan conquered Manchuria and later invaded China. Korea then became a crucial link for the movement of the Japanese forces. Whilst the developed railroads were extensively used, there was a major drive to set up industries right up to the peninsula's northern side. This required relocation of the labour force (and thereby, dislocation of the population) to the areas where the industries were located and, at times, had to be done by force. As the importance of Korea increased and it graduated to becoming a crucial link in the Japanese strategic calculation, there was also a drive to assimilate the Koreans to "become Japanese" and be of utility in war.

From the late nineteenth century to 1945, Korea interacted closely with three different models of governance: the Japanese; Western Europe and the US; and the Soviet Union. Koreans eventually adopted in-parts some of the characteristics of each of these systems. The Korean system imbibed the concept of State and society from the Japanese; liberal democratic ideas from Western Europe and the US; and socialist ideas from the Soviet Union.[13] In the final years of its rule, Japan provided Korea with a model of state-directed economic development, with examples of mass population mobilisation for national purposes and organising massive propaganda campaigns. During the entire period, all major Japanese Zaibatsu (industrial financial conglomerates) operated through the Governor-general in Korea. Parallelly, local entrepreneurs grew after the 1910s and many aristocrats who owned landed property began to enter business. This period also shaped the lives of many prominent personalities, including industrialists Chung Ju-yung (1915-2001) – founder of Hyundai; Lee Byung-chul (1910-1987) – founder of Samsung; leaders such as Park Chung-hee (1917-1979) who was the President of South Korea in crucial periods of reform, and the North Korean dictator Kim Il Sung (1912-1994).

After World War II

Japan surrendered to the Allies on 15 August 1945, and in the wake of the sudden withdrawal of the Japanese from the Korean peninsula, the Americans – under Truman – proposed to divide the area into two zones along the 38th parallel, with the southern zone under their administration. The Soviets – under Stalin – agreed to the proposal and advanced their forces to take control of the northern part.[14] Although the Allies had originally planned that the entirety of Korea would eventually be governed under the joint trusteeship of the US, China, the UK, and the Soviets, subsequent developments resulted in the formation of a democratic government in the South – the Republic of Korea was formed under US influence, with Syngman Rhee as the president. The Soviets subsequently administered through a communist regime in the North, with Kim Il Sung – an ex-guerrilla soldier from the 88th Red Army brigade.

Around 1946, during the time of Korea's division into the North and South, the North had most of the industries and was economically prosperous whereas the South (ROK) was left with limited resources – most of which were in agriculture. In the following years, North Korea – under Soviet influence, implemented major land reforms and organised its economy to build a strong nation with a powerful army, in contrast to the ROK, which continued to lag under poor administration and corrupt governance despite extensive US aid. The North started the Korean War (1950-1953) with Soviet backing and Chinese support, aiming to unify Korea under a single communist regime. However, the US decided to intervene directly, and the war ended with the restoration of the status quo along the 38th parallel – the dividing line between two independent countries.

Syngman Rhee managed to come back to power and continued to be at the helm of affairs till 1960. Though ROK did not make significant economic progress during this period, it rebuilt its infrastructure which had been devastated during the war, using financial aid from the US. The US used its influence to implement land reforms under which an individual could hold property of up to 7.5 acres; farmers who received lands under re-distribution had to pay 150 per cent of the present cost to the government over a ten-year period.[15] The impact of these reforms was that tenancy of land became virtually non-existent and erstwhile peasants became small entrepreneurial farmers. Conservative landlords moved on from agriculture to commerce, industry, and education.[16] In 1958, the government formed an Economic Development Council which was a committee of experts that drew up a detailed plan for the economic growth of the country.

On 16 May 1961, South Korea came under martial law following a bloodless coup which ended with the formation of a Supreme Council for Military Reconstruction (SCNR) under General Park Chung-hee. In the systemic purge that followed, about 40,000 members of the bureaucracy were dismissed; 4000 politicians were prohibited from any political activity for six years; and a revolutionary tribunal was formed under the military, which tried thousands of offenders for corruption and treason.[17] During his formative years, Park Chung-hee was impressed by the state-directed economic development achieved by Japan, and set out to implement a similar model. Whilst his ideas were influenced by revolutionaries like Sun Yat Sen, Kemal Pasha, Nasser, and the Meiji rulers of Japan, the Meiji system's success – in particular – became central to the system he put in place. As per this system, he recognised the importance of learning from and indigenising foreign ideas; establishing a political hierarchy headed by an emperor (a role that he saw for himself); and facilitating millionaires so that they could take centre stage, thereby promoting reforms and encouraging national capitalism.[18]

The SCNR began to implement the recommendations of the Economic Development Council and focussed on developing a self-reliant economy in the first five-year plan (1962-66). Thirteen prominent entrepreneurs were appointed to a Promotional Committee for Economic Reconstruction which advised the government. The first five-year plan encouraged the development of light industries for export and achieved an annual growth rate of 8.9 per cent against the targeted growth rate of 7.1 per cent. During this period, exports grew 29 per cent per year and manufacturing grew by 15 per cent per year.

The second five-year plan (1967-71) focussed on improving infrastructure, transportation, and electric power. Park committed to deploying Korean troops to Vietnam to assist the US forces, and in return, got major construction contracts for Korean companies, a substantial increase in US aid, and access to exports to the US markets. ROK normalised relations with Japan in 1964 and facilitated massive Japanese investments and the transfer of technologies to boost its export-oriented manufacturing industries. The third five-year plan (1972-76) focussed on investing in Heavy and Chemical industries and identified steel, chemical, metal, machine building, shipbuilding, and electronics. This economic development plan aimed to develop shipbuilding as a strategic export industry by increasing shipbuilding capability. The plan was precisely executed and as a result, in the subsequent decade, South Korea emerged as the world's second-largest shipbuilder.[19]

The Korea Shipbuilding and Engineering Corporation (KSEC) was a major government-owned shipyard; the government decided to expand this shipyard to promote shipbuilding.[20] The KSEC catered to the requirements of the merchant marine as well as the Navy. It was established in 1937 by the Japanese Mitsubishi group as a part of the Japanese war policy and was a fully operational shipyard. In the early period after independence, neither the entrepreneurs nor the government had considered shipbuilding as a sector with high growth potential as there was no indigenous steel industry. The capability in shipbuilding was also restricted to building only small ships. However, the government progressively established the Department of Naval Architecture in the universities of Seoul, Pusan, and Inha between 1946 and 1954, and funded a Korean Shipbuilding Society project to develop a series of standard model ship designs. KSEC, Seoul National University, and Pusan National University collaborated as a part of this project and developed sixty standard designs which were then shared with all the shipbuilders of the country.

In the late 1960s, ROK managed to get funds from Japan as a part of the reparations and set up a consultancy group called the 'Japan Group' consisting of Nippon Steel and others to set up a world-class steel production facility. Pohang Iron and Steel Company (POSCO) was set up as an integrated iron and steel company with a production capacity of 9.1 million tons. The first phase was completed by 1972, following which production commenced. Whilst finance and technology came from Japan, the ROK government under Park facilitated growth by providing tax reliefs and assured market demand. POSCO demonstrated the successful model of setting up a world-class heavy industry, managing finance and the know-how from globally established professional companies. Although POSCO was a government-owned company rather than a *chaebol*, the stage was set for ROK to venture into heavy industries as well as government-identified shipbuilding as an industry with great export potential.

The Hyundai group took the lead to venture into shipbuilding. It was one of the largest *chaebols* in ROK, created in 1947 by Chung Ju-yung as primarily a construction company, subsequently diversifying into different fields as it expanded. Initially, shipbuilding was founded as a department in the Hyundai Construction Company; in 1973, it was established as the Hyundai Heavy Industry (HHI), aiming to build ships and heavy machines. The construction of HHI began in 1972, as did the construction of its first ship simultaneously – a 260,000 DWT VLCC in 1973. In about a decade, it grew to be the largest shipyard in the world. The government lobbied extensively with international financial agencies, including the IMF and World Bank, to convince them against their preconceived notions that ROK was not ready to enter the shipbuilding sector. The government prevailed, facilitating investments in terms of overseas credits for the HHI to build the shipyard by providing financial guarantees, both for the shipyard construction as well as enabling initial orders from international companies. The government also provided extensive subsidies for infrastructure.

The HHI identified suitable international sources to tap expertise in world-class shipbuilding. It collaborated with A&P Appledore – a naval architecture firm in Scotland – to get the design of a dockyard, and Scott Lithgow – a Scottish shipbuilding firm – for ship designs and operating instructions. The HHI sent many newly recruited naval architects for training to the premises of A&P Appledore, where the detailed design of the dockyard layout was finalised, and drawings made. The Hyundai Construction Company constructed the dockyard under the supervision of A&P Appledore. Experienced European shipbuilders worked as employees of the HHI for the first three years of operation. The HHI also collaborated with Japan's Kawasaki Shipbuilding Company to learn ship production know-how.[21]

The HHI won its first order from a Greek shipowner George Livanos for building two VLCCs of 260,000 DWT. One of the contract's conditions was that both the ships would be built identically to the ones Scott Lithgow built earlier. Scott Lithgow provided detailed drawings as well as the equipment list to the HHI, enabling them to source most of the major equipment from European companies. The steel plates arrived on site by 1973 when only one of the seven drydocks was completed. Hence, the fabrication and welding activities of shipbuilding commenced parallell to the construction of the shipyard. During this period, the HHI appointed an experienced Danish shipbuilder as the President, employing qualified European shipbuilding professionals in departments like training, production planning, hull construction and production, machinery and electrical installation, etc.

This model stands out as unique, wherein a private enterprise set out to establish a world-class shipbuilding yard, extensively supported by the government in terms of facilitating investments and finances, providing a range of subsidies, and creating a demand from international buyers. Another unique aspect of this venture was its technology strategy, as a part of which the design and layout of the shipyard as well as the ship design capability were acquired from Europe's established shipyards. This capability was enhanced by training a large number of Korean engineers at their premises and also employing several European specialists to work, both in the construction of the shipbuilding yard as well as in ship construction for the first three years. Whilst the HHI acquired its shipyard and ship design from the West, it modelled its production engineering on the lines of the Japanese, acquiring the same from the Kawasaki shipbuilders.

Although both the Europeans and the Japanese provided technical assistance, in European practice, skilled workers wielded greater discretion over their job content and methods. Because the Korean shipyards, like the Japanese before them, were short of experienced skilled workers during their early years of operation, the Japanese practice of a centralised definition of job content and methods was followed instead. The benefit of this carefully formulated technology strategy was that the shipyard evolved, utilising the best practices of European- and Japanese-established shipbuilders, and managed to get initial orders from the erstwhile customers of these very companies which helped them establish themselves in the global shipbuilding market.

Despite the well-thought-through plan and setting up, once the foreign experts left the company, the shipyard began to experience delays in projects and rejection of quality. To address these, the HHI resorted to forward-integrating by starting its own shipping company – Hyundai Merchant Marine – that bought out ships which were not accepted by customers due to delays. To achieve vertical integration, the HHI created in-house design capability, set up a quality department, and refined its production processes by modifying material flow and job sequencing based on time and motion studies. The HHI also set up its own Heavy Machinery Manufacturing Company (HEMCO) which built branded engines under the supervision of its licensors, B&W of Denmark, and MAN, Sulzer, and SEMT Pielstick of Germany.[22]

The technology strategy in each of these initiatives was consistent – infusions of foreign technical assistance; dispatch of numerous trainees overseas; sequential import-substitution of parts and components; and development of local sub-contracting networks and diversification. All these initiatives helped reduce the overall cost of the ship and provided a competitive advantage to the HHI. The HHI, therefore, gained from the entrepreneurial versatility of the private sector (being a family-owned *chaebol*) and solid financial and policy support from the government. Starting off as a greenfield project[23] in the early 1970s, it could outperform the contemporary Japanese shipbuilders based on the economy of scales and lower costs, emerging as the world leader in shipbuilding by the mid-1980s.[24]

The HHI model became a blueprint for other *chaebols* to follow and was the illustration of the rise of shipbuilding in ROK. Since the late 1960s and through the 1970s, the Korean government facilitated the setting up- and expansion- of a world-class export-oriented shipbuilding capability. In 1973, the government formulated a Long-term Shipbuilding Industry Promotion Plan that promoted the setting up of major shipyards, and simultaneously created demand for these shipyards by introducing a Planned Shipbuilding Programme (PSP). Under the PSP, it became mandatory that Korean cargo be shipped only by Korean ships. Shipyards were also provided steel indigenously at reduced prices of the order of 8 per cent so that Korean shipyards could outperform competitors in the global market. Korean shipping companies were promoted through special rights to ship cargo. For example, the owners of bulk carrier ships were given the right to transport imported fertilizers for a duration of five years at higher freight charges. The government-owned banking system played a vital role in providing export credits at low rates. Many industries – which provided input materials for shipbuilding – were built in clusters in shipyards' vicinities to evolve a seamless logistic supply chain.

The main shipyards were getting established in the south-eastern part of the Gyeongnam region in the districts of Ulsan, Busan, and Geoje. In 1978, Daewoo opened its shipyard in Okpo, while Samsung established its yard in Koje in 1979, and STX set up its yard in 2001. Each of these shipyards could construct large-size vessels and had the provision for future expansion to scale up production. The shipyards began production of ships in record time and were soon making large profits from the international shipbuilding market. The profits were routed back to further enhance the infrastructure and capacity of the shipyards. The government also facilitated the setting up of industry clusters that produced shipbuilding material and ancillary industries in regions easily accessible to the shipyards. Most of the input materials and machines required were available locally. The marine equipment cluster of engine manufacturers, steel fabrication units, and other shipbuilding vendors had an estimated export value of US $27.5 billion.[25]

A major advantage accrued from the economy of scales. As there were usually a couple of major shipyards which were part of a single *chaebol*, they leveraged their large capacities and huge order books to optimise on economies of scale while procuring material. For example, the HHI which has 3 shipyards with 9 building docks (BDs) can deliver 70 to 80 large ships in a year, yielding an annual revenue of about US $6 billion. During its peak performance, the shipyard has a firm contract for 200 ships, which translates to three years of work. Each year, the HHI procures items worth US $4 billion from its vendors.[26] It can therefore negotiate for the best price. In addition, a stable forecast of demand for three years' roll-on period makes it easy, both for the HHI as well as the vendors, to plan better and share lessons learnt along with the outcomes of R&D efforts.

Shipbuilding in ROK gained momentum in earnest in the early 1970s, with large *chaebols* taking the lead and the government facilitating through subsidies and loan guarantees. During the early days, global demand for ships was at a record low because of the ongoing world oil crises. The shipyards used the depressed shipping market to gain technologies from established shipbuilders in Europe and Japan, in addition to capital from Japan as part of reparations. Soon after the world oil crises ended, demand for ships grew as the global economy recovered, following which Asia experienced a phase of rapid economic growth. As China's economy began to grow at a fast pace, sea-borne trade between Asia and the US increased, and there was a high demand for large vessels (those that could stack up 10,000 containers on the decks). During this period, a regulation was promulgated by the IMO, calling for the destruction of ships that were older than 25 years. Many shipping companies had to switch over to double-hull tankers to meet the timelines of the strict environmental IMO regimes by 2010.[27] Shipyards thrived on rising demand, advantaged by cheap labour, the low cost of input materials like indigenous steel, etc.

During the slump in global demand, Korean shipyards sustained on orders from shipping companies that were either owned by or partners of *chaebols*. This helped them continue to improve on quality and reduce costs of production. Once the global demand picked up in 1988, the Korean shipyards scaled up their capacity, and by the mid-1990s, Korea had captured a 24 per cent share of the global shipbuilding market. During the ensuing Asian crisis from 1997 to 1999, the Korean economy was adversely affected, and their currency (Won) was devalued against the US dollar. In the same period, however, the Japanese Yen appreciated against the US dollar. The Korean shipbuilders therefore gained from the crisis, as the cost of ships being in US dollars meant that the construction costs in Korea came down. Encashing on the cost advantage, Korea increased its share in global shipbuilding and by 2000, emerged as the world's largest shipbuilder. Around 2006, there were nine major shipbuilding companies capable of building ships larger than 5000 GT and were contributing to around 95 per cent of the shipbuilding output of the country. The balance output was from 54 minor shipyards.

All major shipyards have streamlined their business flow by integrating forward with shipping companies to partly stabilise demand, and have integrated vertically with engine manufacturers and ancillary equipment providers. They have either set up holding companies within their *chaebols* or acquired substantial stakes in these enterprises. The shipyards expand through collaboration with international shipyards or by setting up their own branches in other countries to acquire newer technologies, like the construction of LPG and CNG ships, or setting up offsite shipyards to capitalise on lower labour costs. For example, STX set up STX Europe to acquire technology to construct complex vessels like LNG and LPG carriers. In order to capitalise on lower labour costs, STX established a shipyard in Dalian, China, and SMD did so in Vietnam. Shipyards create expertise in building all types of ships rather than verticalising one type. They also diversify into other products so that they can sustain when the demand for shipbuilding is low.

The Warship-Building Industry After World War II

Naval shipbuilding in ROK is embedded in the much larger, commercially-oriented shipyards which also specialise in building modern weapons-intensive capital warships and submarines. Naval shipbuilding, therefore, has been able to take advantage of infrastructure like drydocks, production technologies like CAD/CAM, and a massive pool of skilled labour for export-driven merchant ship production. Destroyers, Frigates, and Corvettes are built in the HHI and Daewoo shipyards. In addition, the Korean Tacoma and Korean SEC yards also build frigates and corvettes. The submarines and coast guard ships were also constructed by the HHI and Daewoo. The Landing Platform Docks (LPDs) have been built by the HHIC. Since different shipyards build the same class of ships, the Korean Navy has the flexibility to choose from amongst the shipyards and therefore take advantage of competition. The modernisation of ROKN relies increasingly on indigenous sources for both the platforms as well as the systems.[28]

One of the areas that underscores the technological sophistication of the naval shipbuilding industry of ROK is its indigenous submarine-building capability. Daewoo Shipbuilding and Marine Engineering (DSME) and Hyundai Heavy Industries (HHI) began building Type 209 diesel-electric attack submarines under licensed production from HDW, Germany in the 1990s. They built 9 submarines in the KSS-I or *Jang Bogo*-I class. In the 2000s, the KSS-I class was superseded by the KSS-II/*Son Won*-II class which was Type 214 under license production from the same shipyard. Between 2006 and 2017, nine KSS-II class submarines were launched. These were hybrid diesel-electric/ fuel cell submarines with Air Independent Propulsion (AIP) technology.[29] The submarine-building capability of ROK evolved to an extent wherein all submarines of the ROKN were indigenously built. ROK earned a reputation as an international submarine builder; by 2003, ROK shipbuilders were carrying out engine replacements of the Indonesian type-209 submarines. In 2011, ROK shipbuilders outbid Russia, France, and Germany on a $1.1 billion tender to supply three type 209s to Indonesia.[30]

The submarine capability has continued to grow and a 3000-ton diesel-electric submarine christened *Dosan Ahn Changho*, a KSS-III class, was launched in September 2019, and was slated to be commissioned by 2021. The submarine is equipped with a fuel cell air-independent propulsion system, conventional torpedo tubes, and a vertical launch system for launching anti-ship and land attack cruise missiles.[31] Four KSS-IIIs are already under construction and they use homegrown technologies, once again demonstrating ROK's capability to effectively absorb foreign technology through licensed production and successfully grow indigenous capability to compete globally. As the technologies of Industry 4.0 like AI, big data, and autonomous operation mature globally, the ROKN is implementing the vision of a "Smart Navy" under the "Defence Reforms 2.0" programme. The programme prioritises the development of Unmanned Underwater Vehicles (UUVs) and extra-large UUVs (XLUUVs) which could use the KSS-class submarines as a mothership against sea mines, submarines, and SLBMs.[32]

The naval surface shipbuilding has also developed modern warships with contemporary designs as per a structured long-term programme, like the

Korean Destroyer Experimental (KDX). These ships were designed and constructed in three phases, starting with KDX-I in 1998 and were made in the DSME or the HHI. The KDX-III (*Sejong the Great*-class), destroyers with the most advanced design, are identical in design to the *Arleigh Burke*-class destroyers of the USN. They are equipped with the combined gas turbine (COGAG) propulsion system, Aegis weapons system, and a range of indigenous missile systems.[33] In 2018, the process of indigenously designing the Korean Destroyer next Generation (KDDX) was initiated, with the aim of finalising the design and inducting the ships into the ROKN by the late 2020s. In March 2020, General Electric and Hanwa Aerospace signed an MoU with the HHI and DSME shipyards for electric propulsion systems in the KDDX programme – which will largely have the most advanced, indigenously-made weapons and missile systems.[34] Under the exports programme, the South Korean shipyards have sold two frigates to the Philippines and LPDs to Indonesia, the Philippines, and Peru.[35]

Recent Developments in Shipbuilding

The South Korean shipbuilding industry not only became the largest shipbuilders in the world in just three decades, but they have also maintained their position in the top two ever since. In order to maintain their global leadership, the shipyards have invested heavily in green technologies and the emerging applications of Industry 4.0, like AI, VR/AR, Autonomous vessels, IoT, and blockchain. The technologies of Industry 4.0 are bringing about a paradigm shift, where they bring in greater efficiency in shipbuilding in addition to enabling the shipbuilders to remain connected with the shipowners, by providing features to economise ship operations and reduce ownership cost. Shipyards like the HHI are using robotics and AI extensively as they graduate to smart manufacturing. The use of a robotic system which has a high-frequency inductive system, a multi-joint arm based on automation technologies, and IoT to automatically shape a large vessel's 3D curve surface is an excellent example of smart shipbuilding.[36]

In order to consolidate the global lead of ROK, the Ministry of Trade, Industry and Energy *and* the Ministry of Maritime Affairs and Fisheries have together established a task force to coordinate efforts by different agencies in the fields of AI, big data, and IoT for the development of autonomous vessels.[37]

The HHI has developed and successfully deployed applications like the Hyundai Intelligent Equipment Management System (HiEMS) that collects data from the engines and controls during operations, applies AI in real-time, and makes adjustments to optimise fuel efficiency. The system claims to have achieved as much as 10 per cent savings in fuel consumption.[38] The Hyundai Intelligent Navigation Assistance System (HiNAS) uses radar, including ARPA, AIS, and data from forward-looking cameras; and applies deep learning, sensor fusion, and Augmented Reality (AR) to assist the crew in navigation safety.[39]

The HiNAS and HiBAS (Hyundai Intelligent Berthing System) are core technologies of autonomous ships, and some of them are already installed onboard ships like the 250,000-ton bulk carrier of SK Shipping.[40] The HHI is installing IoT sensors during construction and then providing IoT-based applications, developed by them under an MoU with SK Shipping, Intel, Microsoft, and the Centres for Creative Innovation Economy and Innovation at Ulsan and Daejeon (UCCEI and DCCEI). These IoT applications will enable ballast tank inspection, remote medical treatment of the crew members, automatic reporting of voyage information, maintenance support and safety monitoring, etc.[41] All these applications add up to building 'smart' vessels. In 2019, the American Bureau of Shipping (ABS) certified the HHI's integrated smart ship solution.

Korea is the first country to make commercial 5G available in the world. In November 2019, the HHI and KT signed an MoU to work on creating 5G-powered 'smart shipyards' which could use wearables like neckband cameras that provide data in real-time to an integrated control tower.[42] The DSME delivered a 24,000 TEU class container ship to HMM, equipped with the DSME smart ship platform called DS4®. This platform will be fitted on the next seven ships of the same class and will enable the shipowners to remotely diagnose major systems such as main engines, air conditioning systems (HVAC), and refrigeration containers to support maintenance. These ships also have smart navigation and other smart platform systems with require layers of cyber security. They also meet the latest IMO environmental and energy standards.[43]

Samsung Heavy Industries (SHI) has developed a cloud-based smart ship system (SVESSEL), in collaboration with MAN, to include enhanced operating services for main engines, diagnostics and controls. It has a similar collaboration with WinGD of Switzerland for LNG-fuelled vessels. This is a new trend where shipbuilders are collaborating with equipment producers for next-gen innovative smart ship systems.[44] At the end of 2019, Samsung conducted trials of a 5G-based autonomous and remotely-controlled navigation platform. The system was developed jointly by Samsung and SK Telecom and was fitted on a shuttle tanker which received DNV GL's 'Smart ship' notation.

NOTES

1   "South Korea - Economic and Social Developments," *Encyclopedia Britannica*, https://www.britannica.com/place/South-Korea, accessed 16 May 2020.

2   A slightly larger-scale form of gardening adopted by a farmer family to grow crops in the small individual plots of land allotted to them by the state. Agriculture was restructured by land reforms under which large landholdings were taken over by the government and redistributed in small plots to individual farmers for highly labour-intensive household farming. This maximised the yield from agriculture, resulted in distributed capital gains to a large percentage of the population, and led to the productive surplus that primed demand for goods and services.

3   Sectors that enable capital accumulation, rise of series of linked industries, and facilitate innovation and growth of many sectors.

4   Kyoung-Ho Shin and Paul Ciccantell, "The Steel and Shipbuilding Industries of South Korea: Rising East Asia and Globalization," *Journal of World-Systems Research* vol. 15, no. 2 (2009), 167–192.

5   MJ Seth, *A History of Korea: From Antiquity to the Present* (Rowman & Littlefield, 2010), 6–7.

6   JB Palais, "A Search for Korean Uniqueness," *Harvard Journal of Asiatic Studies* vol. 55, no. 2 (1995), 409–425.

7   James B. Palais, *Politics and Policy in Traditional Korea*, (Cambridge: Harvard University Press, 1975), 4–5.

8   William Elliot Griffis, *Corea the Hermit Nation: I.—Ancient and Mediaeval History. II.— Political and Social Corea. III—Modern and Recent History* (New York: C. Scribner's sons, 1882), 7–9.

9   Key Ray Chong, "The Tonghak Rebellion: Harbinger of Korean Nationalism", *Journal of Korean Studies, (1969-1971)* vol. 1, no. 1 (1969), 73–88.

10 CK Quinones, "The Impact of the Kabo Reforms upon Political Role Allocation in Late Yi Korea, 1884-1902," *Occasional Papers on Korea*, no. 4 (September 1975), 1–18.

11 Bruce Cumings, *Korea's Place in the Sun: A Modern History (Updated Edition)*, (New York: W. W. Norton & Company, 2005), 105–15.

12 Alice H Amsden, *Asia's Next Giant: South Korea and Late Industrialization*, (New York: Oxford University Press, 1989), 28–30.

13 Seth, *A History of Korea.*, ch.6, 254–56.

14 Jongsoo James Lee, *The Partition of Korea After World War II: A Global History* (New York: Palgrave Macmillan, 2006), 38–42.

15 Ki Hyuk Pak, "Outcome of Land Reform in the Republic of Korea," *Journal of Farm*

Economics vol. 38, no. 4 (1956), 1015–23.

16 John Lie, *Han Unbound: The Political Economy of South Korea* (Stanford: Stanford University Press, 1998), 9–18.

17 Joungwon Alexander Kim, *Divided Korea: The Politics of Development, 1945-1972* (Elizabeth, NJ: Hollym Intl, 1999), 229–33.

18 Chung Hee Park, *The Country, the Revolution and I* (Seoul: Hangchun-Dan Mansion, 1963), 14–16.

19 Seth, *A History of Korea.*, ch.6, 380-90.

20 Hourtouat Florence, "Peer Review of the Korean Shipbuilding Industry and Related Government Policies," *OECD Council Working Party on Shipbuilding Report C/WP6(2014)10/Final* (Paris: OECD, 2015), 63, https://www.oecd.org/officialdocuments/publicdisplaydocumentpdf/?cote=c/wp6(2014)10 /final&doclanguage=en, accessed on 28 June 2020.

21 Amsden, *Asia's Next Giant,* ch.1, 276–77.

22 Amsden, *Asia's Next Giant,* 280.

23 A greenfield project starts ab initio on a barren site, hence there is no requirement for demolition or remodelling of any existing facility as against a brownfield project that involves modifications to the existing facility.

24 Henry Wai-chung Yeung, *Strategic Coupling: East Asian Industrial Transformation in the New Global Economy* (New York: Cornell University Press, 2016), 123.

25 Krishnan, *Prosperous Nation Building Through Shipbuilding,* ch.2, 116–18.

26 Krishnan, *Prosperous Nation Building Through Shipbuilding,* ch.2, 113–15.

27 Shin and Ciccantell, *The Steel and Shipbuilding Industries of South Korea,* ch.6.

28 Ian Bowers, *The Modernisation of the Republic of Korea Navy: Seapower, Strategy and Politics* (New York: Springer, 2018), 84.

29 An advanced propulsion system that uses fuel cells to enable the submarine to remain underwater for a longer duration.

30 "South Korea Submarine Capabilities,", https://www.nti.org/analysis/articles/south-korea-submarine-capabilities/, accessed 11 July 2020.

31 RA Bitzinger, "S Korean Naval Shipbuilding: Full Speed Ahead," *Asia Times,* https://asiatimes.com/2019 /07/s-korean-naval-shipbuilding-full-speed-ahead/, accessed 30 June 2019.

32 Sukjoon Yoon, "Make Way for South Korea's Underwater Drones," *The Diplomat,* https://thediplomat.com/2020/02/make-way-for-south-koreas-underwater-drones, accessed 19 February 2020.

33 "Sejong the Great Class / KDX-III Class Destroyer," *Naval Technology,* https://www.naval-technology.com/projects/sejongthegreatclassd/, accessed 11 July 2020.

34 "South Korea Invites Local Firms to Design Aegis-Equipped Destroyer," https://www.defenseworld.net/news/27102/South_Korea_Invites_Local_Firms_to_Design_Aegis_equipped_Destroyer#.XwnWbSgzY2x, accessed 11 July 2020.

35 Bitzinger, *op cit.,* ch. 6, n. 30.

36 Leah Alger, "Hyundai HI to Use Robots for Shipbuilding," *Software Testing News,* https://www.softwaretestingnews.co.uk/hyundai-hi-to-use-robots-for-shipbuilding/, accessed 15 May 2018.

37 "South Korea Coordinates Autonomous Ship Efforts," *The Maritime Executive,* https://

www.maritime-executive.com/article/south-korea-coordinates-autonomous-ship-efforts, accessed 17 June 2020.

38 Jo He-rim, "Hyundai Heavy Industries Uses AI to Create 'Smart' Vessels," *The Korea Herald*, http://www.koreaherald.com/view.php?ud=20200113000667, accessed 13 January 2020.

39 Maro Jeon, Jinmo Park, and Joohyun Woo, "Development of HHI's Advanced Navigation Assistance System for Safe Voyage," *12th IFAC Conference on Control Applications in Marine Systems, Robotics, and Vehicles CAMS 2019*, vol. 52, no. 21 (2019): 111–13.

40 Chang-won Lim, "Hyundai Shipyard Applies Autonomous Sailing Technology to Bulk Carrier," *Aju Business Daily*, http://www.ajudaily.com/view/20200409173629074, accessed 09 April 2020.

41 "Shipbuilder Looks to Internet of Things for Future Business," *The Maritime Executive*, https://www.maritime-executive.com/article/shipbuilder-looks-to-internet-of-things-for-future-business, accessed 07 June 2016.

42 James Jung, "Shipbuilding to Benefit from KT, Hyundai Heavy's 5G Tech," *Korea Tech Today*, https://www.koreatechtoday.com/shipbuilding-to-benefit-from-kt-hyundai-heavys-5g-tech/, accessed 16 December 2019.

43 MI News Network, "DSME Solidifies Position As The World's Leading Smart Ship Building Leader," *Marine Insight*, https://www.marineinsight.com/shipping-news/dsme-solidifies-position-as-the-worlds-leading-smart-ship-building-leader/, accessed 22 May 2020.

44 World Maritime News, "Samsung Heavy Joins Forces with MAN on Smart Ship Tech," *Offshore Energy*, https://www.offshore-energy.biz/samsung-heavy-joins-forces-with-man-on-smart-ship-tech/, accessed 19 August 2019.

7

Case Study – China

Introduction

China is an outstanding example of economic transformation in recent history and is unmatched, especially in the pace and scale of such a transformation. It is all the more significant considering that it is one of the largest and most populous countries. China's per capita GDP has grown at a yearly average in excess of 8 per cent per year ever since 1978, and from being one of the poorest countries in the world, it is now the second largest economy after the US.

The Chinese shipbuilding and ship repair industry comprises yards big and small, coastal and inland, with the primary shipbuilding and ship repair activities concentrated in Shanghai, Guangzhou, and Dalian. These activities have developed at the mouth of the Yangtze and Pearl Rivers, with limited development on the eastern coastline between these two rivers and in the coastal areas bordering the Bohai Gulf and the mouth of the Yellow River in the north. The largest shipbuilding cluster in the country is in the Yangtse River Delta region, having shipyards in the provinces of Shanghai, Jiangsu, Anhui, and Hubei.[1]

The Pearl River is the longest river in South China and is home to shipbuilding facilities around the Guangdong, Guangxi, Guizhou, and Yunnan provinces. In addition to this conglomerate, there are other shipyards, including those owned by China Ocean Shipping Company (COSCO), Provinces, and the PLAN. Chinese naval shipbuilding has been used judiciously to ensure support to its industry to get sufficient orders to sustain in the face of stiff competition from Japan and Korea.[2] The shipyards in China form a part of the larger marine ecosystem that has evolved from the historical maritime lineage of the country.

History Before the Mid-Nineteenth Century

China is one of the earliest civilisations that emerged in the plains of the Yellow River (to its north). It was ruled by successive dynasties, starting with the Xia Dynasty as early as 2070 BCE, when small villages and farming communities worked together to control the flooding of the Yellow River and evolved a centralised government. Ever since then, China has disintegrated, re-united, and expanded multiple times under numerous successive dynasties, including the Shang, Zhou, Qin, Han, Xin, Tang, and others. This continued until 1912 when the Xinhai Revolution overthrew the Qing Dynasty and established a republic. Many of the dynastic periods marked high points of Chinese civilisation. From 206 BC to 220 AD, the Han Dynasty marked the advent of advanced technologies, such as papermaking and the use of the compass, and improvements in the fields of agriculture and medicine; the Tang Dynasty (618 to 907 AD) is credited with the invention of gunpowder and extending Chinese trade links up to Mesopotamia and the Horn of Africa through the Silk Route.

The Chinese Empire underwent multiple reconstitutions under different dynasties in its last millennium of imperial history, unlike empires in other parts of the world, including the Mauryan (322-184 BCE), Gupta (320-550 CE), and Mughal (1526-1857) Empires in the Indian subcontinent, or the Roman Empire in Europe, which experienced more sporadic changes. A common feature across the different dynasties' reigns in China was that they had a civil-oriented bureaucracy formed by scholars who qualified for their official positions by passing rigorous examinations. Chinese society was dominated by Confucian beliefs and held together – at the most basic level – by a common Han Chinese culture. However, under every dynasty, China was an inherently imperial entity, defined politically and enforced militarily. All dynasties – the Chinese, Mongols, and Manchus – were highly proficient in using warfare for State formation and maintained control through centralised military means under a single ruler, rather than relying on a unified cultural orientation to political order or ethnic unity. This characteristic made them inherently imperial- and conquest-dynasties.[3]

Since military conquest led to the formation of every new dynasty and its effectiveness to reign the State decided its tenure, advancements in military capabilities and the art of war evolved consistently. Considering that China's primary threats came from the nomadic steppe peoples from Central Asia, it has generally been regarded as a country with a continental mindset, having a predisposition towards terrestrial warfare. However, many studies reveal that throughout Chinese history, elements of naval warfare played a crucial role in the creation and unification of the Chinese Empire for over two thousand years.[4] China has a significant naval heritage and there are references to the use of ships in military operations in China as early as 1045 BC. The earliest examples of shipbuilding – dedicated to naval warfare – come from the Spring and the Autumn Period (722-481 BC) when ships were built multi-decked, with rams in the bow, having boarding troops and missile weapons. The Song Dynasty's naval programmes, which began in 960 AD, included early gunpowder weapons as part of their naval armament. They also developed the world's first compass, variable depth rudders for operating in shallow waters, and naval architecture based on watertight bulkheads.[5]

In pre-modern China, many engagements took place on inland waters and were joint operations between armies and navies. Naval operations encompassed – in addition to ship-to-ship combat – the defence of bridges; the transport of invasion forces; riverine and canal patrols; and battles on China's numerous and large lakes. Naval power was centred closer home and – unlike the West – did not focus much on sea-going capabilities. Chinese naval architecture traditionally focussed on vessels of shallower draught with lower beam-to-length ratios and innovative designs. Although these do not fit in the traditional and Mahanian definitions of proper sea battle, naval forces played a decisive role in the transfer of power between Chinese dynasties. Naval river units under provincial command patrolled rivers and canals, forming an elaborate system of hydraulic defence against the nomadic cavalry of the steppe.

Chinese fleets were composed of numerous ships like galleys, transport ships, supply vessels, and those specifically designed for ship-to-ship combat. The Chinese used 'tower ships' to assist in siege warfare. A large tower made up a part of the vessel's superstructure and served as a sea-borne sea tower. The ship sailed to a fortress's walls and troops either fired at the defenders from the superstructure or assaulted the walls directly. These ships are believed to have been used as early as 500 BC and continued to be used up to the fourteenth century. The largest of tower ships are said to have carried two to three thousand troops each. They had numerous decks, watertight hatches, specific holding areas for horses of cavalry units, and armoured superstructure. The tower ships were used to capture cities by scaling their riverine walls directly from the stern of the ships.

In the fifteenth century – from 1405 to 1433 – the Chinese under the Ming Dynasty, put to sea a fleet of 255 vessels with over 27,000 soldiers, sailors, and marines. This fleet, under the command of imperial eunuch Admiral Zheng He, undertook seven voyages – which were part diplomatic missions, part trade missions, and part military force – that patrolled the oceans of Asia, cruising as far as the Persian Gulf and east coast of Africa. The fleet centred around 62 'treasure ships', which featured watertight bulkheads and balanced lug sails. They were 440 feet long, 180 feet wide, and rigged with nine masts – one of the most advanced ship designs in the world.[6] Similarly, Chinese 'junks' were classic sailing vessels of ancient origin but continued to be in use. These vessels were high-sterned with a projecting bow, carrying up to five masts, with square sails consisting of panels of linen or matting flattened by bamboo strips. The hull was partitioned by solid bulkheads, running both transversely and longitudinally, thereby adding greatly to the strength. These Chinese junks were as large and powerful as the European ships of the sixteenth century and could transport a large number of armed men.[7]

China turned its back on the world economy in the early fifteenth century when its maritime technology was superior to that of Europe. This self-imposed isolation left China out of the technological progress made in the West. Colonial ingress which, among other things, enforced unfavourable tariffs in trade, further accentuating China's slowdown.

Mid-Nineteenth Century to the End of World War II

By the mid-nineteenth century, the Manchu Dynasty was in a state of collapse. After its eventual collapse, the Kuomintang Regime that followed was equally incompetent. Colonial ingress commenced with the capture of Hong Kong by the British in 1842, guaranteeing free access to Canton to exchange Indian opium for Chinese tea. A second Anglo-French attack in 1858-1860 opened access to interior China via the Yangtse, where the huge network of internal waterways debouched at Shanghai. Subsequently, free trade treaties were imposed on China to permit trade with European countries, Japan, the United States, and three Latin American countries on a most-favoured-nation basis. The treaties forced China to maintain low tariffs, legalised opium trade and allowed traders from foreign countries to travel and trade in 92 of the 'treaty ports' which were opened between 1842 and 1917.

A multilateral colonial regime controlled the collection of tariffs as revenue. This regime was manned by the citizens of Britain, France, Germany, Italy, Japan, and the USA, and operated exclusive international settlements in Shanghai and some leased territories adjacent to Hong Kong. Foreign trading companies were the primary beneficiaries of imposed free trade imperialism, while the native economy suffered. Major intrusions on Chinese sovereignty and damage to its economy came from Japan, with military incursions intensifying during the Meiji Period after the 1870s. Japan established its suzerainty over Korea in 1876, Taiwan in 1895, and later, southern Manchuria in 1905. China regained its tariff autonomy from the foreign powers in 1928, with some relaxation of constraints on its sovereignty at treaty ports. However, complete freedom from the Japanese was only achieved at the end of the Civil War in 1949.[8]

After World War II

The People's Republic of China (PRC) was founded in 1949 after the Civil War, and the Communist Party came to power with Mao Zedong at the helm of affairs. The economy was brought under state ownership and control and grew at a slow rate, largely driven by the increase in inputs of physical and human capital. Although the establishment of the Jiangnan Manufacturing Bureau in Shanghai in 1865 marked the beginning of modern Chinese shipbuilding, foreign shipping companies dominated maritime affairs in China until 1949.[9] Major government initiatives, such as the Great Leap Forward and the Cultural Revolution, did not yield the intended results; instead, they had disastrous consequences on the country's growth. During the period between 1949 and 1978, there was little contact with the outside world. China faced comprehensive embargoes on trade, travel, and financial transactions from the US between 1952 and 1973, and from the USSR from 1960 onwards.[10] China did not have an international merchant fleet and possessed small shipyards which primarily concentrated on the internal market.

In 1950, the Shipbuilding Industry Bureau was formed under the Ministry of Heavy Industry, with the aim of rebuilding the shipbuilding industry. In 1952, the Bureau was placed under the First Ministry of Machine Building. During the first five-year plan (1953-57) the basic policy was to rebuild and expand China's shipyards with Soviet technical and financial assistance. A Sino-Soviet shipbuilding company was established in 1952 in Dalian, where Soviet ships were repaired in exchange for technical assistance and Soviet equipment supplied for the yard's reconstruction. Demand for merchant ships in China was not strong in the 1950s on two accounts. First, until the Sino-Soviet split at the end of the decade, the Soviet Union was China's dominant trading partner, and the goods were exchanged primarily by rail. Second, after the Korean War, the US-led trade embargo prohibited China from using Chinese-flag ships for trading with non-Soviet countries. Chinese shipyards primarily focussed on building naval ships for the PLAN, while merchant ships' demand was met by chartering foreign-flag vessels.[11] Soviet assistance was stopped after the ties were severed in 1960, and the shipbuilding industry suffered a phase of turmoil.

In the early 1960s, global trade grew rapidly in the post-World War global economy. There was a concomitant rise in the demand for new ships. When the Suez Canal closed in 1967, the global prices of all types of ocean-going vessels tripled until 1974. The global order book for the construction of new ships increased nearly five times — from 55 million deadweight tons (DWT) in 1967 to 258 million DWT in 1974. After the 1973 world oil crisis, there was a global economic recession wherein the demand for new ships collapsed, causing massive tonnage oversupply in the international shipping market. By 1978, the global order book for new constructions was virtually back to the levels in 1967. During this collapse, Japan and Western Europe were left with new state-of-the-art facilities and an expanded but redundant workforce. As per some estimates, the capacity of shipbuilding was twice the global demand. Japan and the Western European shipbuilding countries undertook restructuring programmes to reduce capacity. Meanwhile, South Korea entered global shipbuilding, expecting the markets to recover in the 1980s. China emerged as a new entrant to compete in global shipbuilding, in addition to Taiwan and Brazil.

Deng Xiaoping became the General Secretary of the Communist Party in 1977 and initiated the reform process to bring in a socialist market economy. The aim of national revitalisation and economic liberalisation was to make China the world's leading manufacturer. The government decentralised the decision-making process and opened the market to the outside world. Six defence industries, viz. nuclear, ordnance, aviation, space, electronics, and shipbuilding were brought under the control of the state council. The China State Shipbuilding Corporation (CSSC) was set up in 1982 under the state council as an authority to preside over the entirety of civilian shipbuilding in China, which included 26 shipyards; 66 factories and equipment plants; 33 research and development units; and three institutions of higher education with a workforce of about 300,000.

The CSSC had evolved from the traditional machine-building ministry, which was greatly influenced by the military, into a predominantly civilian-run, profit-seeking corporation, albeit under government control. As part of defence conversion, the shipbuilding industry diversified into commercial shipbuilding, particularly in international sales. Its sustained access to foreign equipment, materials, and technical expertise allowed it to prosper and modernise. Major shipyards under the CSSC, like Dalian, Jiangnan, Hudong, and Guangzhou also built warships, although the PLAN continued to manage some shipyards that were directly under the Central Military Commission.[12]

During this period, China adopted a 'two-track approach' to develop ancillary industries; it reverse-engineered the available Soviet equipment and gained access to Western technologies through the Foreign Military Sales (FMS) route. Following the Tiananmen Square crackdown in 1989, China faced complete isolation from the West; even ongoing projects were discontinued. However, the disintegration of the USSR in 1991 and the cash-strapped economic situation of the newly formed Central Asian Republics opened up an opportunity – importing technology from the Russians and unfinished shipbuilding projects from the Ukrainians.[13]

A shipyard requires an ecosystem of associated feeder industries which cater to the inputs to progress with ship construction. These include human resources, land, energy sources, transportation, and ancillary equipment producers and suppliers. The Chinese shipbuilding industry, realising the importance of such a cluster, relocated the existing associated manufacturers as well as shipyards so developed since 1990 in their proximity. An example of such a cluster is Dalian, which was established in 2007. It consisted of Dalian Shipbuilding Heavy Industry at the first level; supported by three associated manufacturers – the Dalian Maritime Diesel Engine Company, the Dalian Maritime Propeller Company, and the Dalian Maritime Valve Company – in the second level; and another 120 sub-factories, 37 ship repair yards, and 15 maritime logistics companies supplying to the second-level manufacturers. Similarly, the Jiangnan Shipyard group was relocated to Changxing Island, along with a whole new shipbuilding base in 2009.

Noticeable changes began to occur in the shipbuilding industry by the 1990s, largely due to the impact of a strong national economy, the development of technology, and support from the government in promoting business opportunities for shipyards. Large investments were also made in research and development to increase the sophistication of the ships being built. The government facilitated demand-side by ensuring that only Chinese-made ships were used for the transportation of goods within Chinese ports; once the building capacity of the shipyards increased, they competed for export orders. For instance, when China Petrochemical Corporation (Sinopec) wanted to import LNG from Australia, it sought LNG carriers that were ordered from Chinese shipyards. This forced the Chinese shipbuilding industry to acquire technology through technological support and supervision from countries like Japan and France in 2008. Subsequently, the industry bagged international orders for LNG carriers from companies including CNOOC Energy Technology and Services, China LNG Shipping, Teekay Energy Partners, and British Gas Services, from 2013 onwards.[14]

At the time of the CSSC's creation, there was severe competition in the international shipbuilding market, and the Chinese government aimed to achieve 'economic reorganisation and administrative streamlining' of the shipbuilding industry. During the mid-1980s, the orders for warships were low as the PLAN was being restructured as well. The main source for the new merchant ship orders – China Oceangoing Shipping Corporation (COSCO) – aimed to rely on the domestic market while promoting exports and focus on shipbuilding while diversifying production. Although there was a constant need to expand the merchant fleet, COSCO continued to place orders on foreign shipyards till the mid-1980s – primarily from South Korea and Japan – in order to take maximum advantage of the rock-bottom prices and convenient finance schemes offered by the host countries.

In the absence of any preferential treatment in the awarding of orders for new constructions, the CSSC ventured to compete in international shipbuilding. It focussed on getting new construction orders for higher-end vessels such as car carriers, refrigerator-container ships, and product chemical tankers, which was at variance with its original aim of exporting basic ships such as bulk carriers and oil tankers. In early 1985, Dalian Shipyard won major new construction orders from Norway, marking the beginning of the CSSC's entry into the high-end export market. It was, however, facilitated by arranging a loan on liberal terms from the Bank of China, agreeing to import around 50 per cent of equipment and raw materials from Europe. The ship was constructed based on a Norwegian design, with a large number of contracts awarded to contractors from Norway.

Over the years, the Chinese evolved advanced shipbuilding methods by incorporating foreign technology. They acquired the capability to build large vessels in drydocks, advanced hull section construction and outfitting, mechanisation, and assembly line work to reduce build times, along with the use of CAD/CAM in design. They could execute new building orders for an LNG carrier at 147,000 m³ and for an FPSO at 300,000 DWT in 2006.[15] The Chinese procured their VLCC design from the Korea Marine Technology Consulting Company of South Korea in 1999, thus becoming qualified to sign orders for building VLCCs for Iran.[16] Another example is the construction of cruise liners, which the Chinese shipbuilder Shanghai Waigaoqiao Shipbuilding Co. is currently building in collaboration with Italy-based Fincantieri for a Hong Kong-based buyer, which will eventually improve many sectors of the domestic shipbuilding ecosystem and global supply chain.[17]

Starting as early entrants primarily serving assemblers and following imported designs for new shipbuilding orders for export, Chinese shipyards expanded to handle an increasing number of domestic orders from the mid-

1980s as global markets began to recover. The Chinese shipbuilding industry grew into a large, geographically dispersed, increasingly modern sector that was at the nexus of a burgeoning civil economy and its defence industrial complex. In comparison, shipbuilding employment in Western Europe declined by 62 per cent and the available shipbuilding capacity of shipbuilding yards decreased by 64 per cent from 1975 to 1990. In China, on the other hand, exports accounted for over 50 per cent of total Chinese ship production on average since the mid-1980s, with ships becoming one of the largest foreign exchange earners in the entire machinery and electronics sector by the mid-1990s, averaging about US $1.5 billion annually.[18]

The CSSC existed as a single entity until July 1999, when it was divided into two distinct entities: the China State Shipbuilding Corporation (CSSC) and the China Shipbuilding Industry Corporation (CSIC). Both these successor entities are state-owned enterprises and report to the state council. These were created with the aim of introducing competition amongst the defence-industrial enterprises to improve quality and efficiency. Both these entities build multiple types of merchant ships and warships. They compete for domestic and international customers. The CSSC and CSIC constitute about 60 to 70 per cent of the total shipbuilding capacity of China, and the remainder 30 to 40 per cent comprise shipyards like those owned by provinces including Fujian, Guangzhou, and Jiangsu; shipyards owned by COSCO and the PLAN; joint venture shipyards like two of the Nantong-Kawasaki-COSCO Shipyards (Japan), Shanghai Edward Shipyard (Germany), Yantai Raffles Shipyard (Singapore), and Samsung-Ningbo Shipyard (Japan).

The CSSC operates shipbuilding and related facilities in eastern and southern China – primarily the facilities located in Shanghai and the provinces of Guangdong and Jiangxi. It controls 58 enterprises – including shipyards, R&D institutes, factories, and various shareholding companies. It employs approximately 95,000 employees and is valued at US $771 million. The CSIC is responsible for shipbuilding and related facilities in north-eastern China – which includes the Tianjin, Hebei, and Liaoning provinces – and inland China – which includes Sichuan and Shanxi Provinces and Shandong Province. The CSIC is a far larger industrial entity – employing about 170,000 employees covering 38 industrial enterprises, 10 shipyards, 28 R&D institutes, and

15 shareholding companies – spread over 20 provinces of China and is valued at US $3.5 billion.

In 2006, China identified shipbuilding as a 'strategic industry' (as part of the eleventh five-year plan of 2006-10) and introduced plans to develop the country's shipbuilding industry to make it the largest in the world. The policies set specific output and capacity goals and covered production, investments, and firm entry subsidies. Production subsidies included input material and export credits, buyer financing, etc.; investment subsidies included low-interest long-term loans. The cost of entry for potential shipyards was lowered by reducing processing time, simplifying licensing procedures, and subsidising land prices. The government invested 20 billion Yuan to establish large production bases in Waigaoqiao, Changxing, and Longxue. Within a short span of time, China's market share of global shipbuilding doubled from 25 per cent to 50 per cent. Although China was a late maritime Asian entrant in global shipbuilding, by the time its shipbuilding industry began to consolidate its global position, the two leading shipbuilders in the world were both from the same region and had captured lead positions in global shipbuilding. A comparison of the global shipbuilding market share among these three countries in maritime Asia illustrates the rapid growth of Chinese shipbuilding, relative to Japan and South Korea. The market share of these leading shipbuilding countries in global shipbuilding is depicted in Figures 7.1 and 7.2.[19]

Figure 7.1: Market Shares of China, Japan, and South Korea[20]

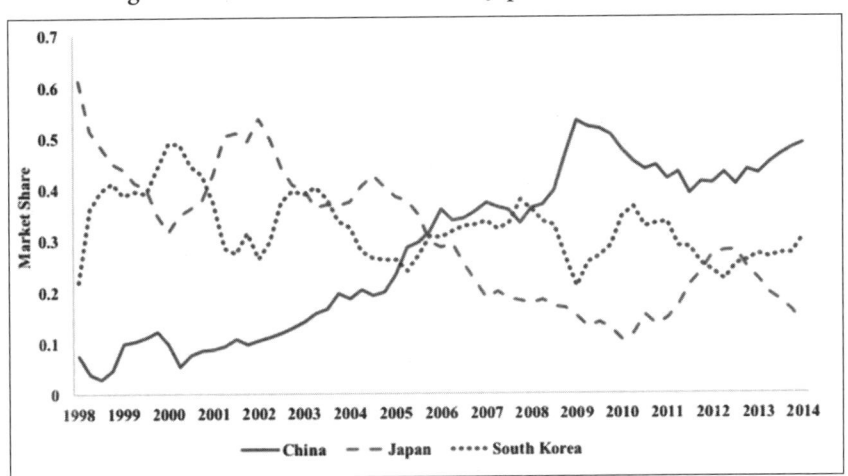

Figure 7.2: Number of Shipyards in China, Japan, and South Korea[21]

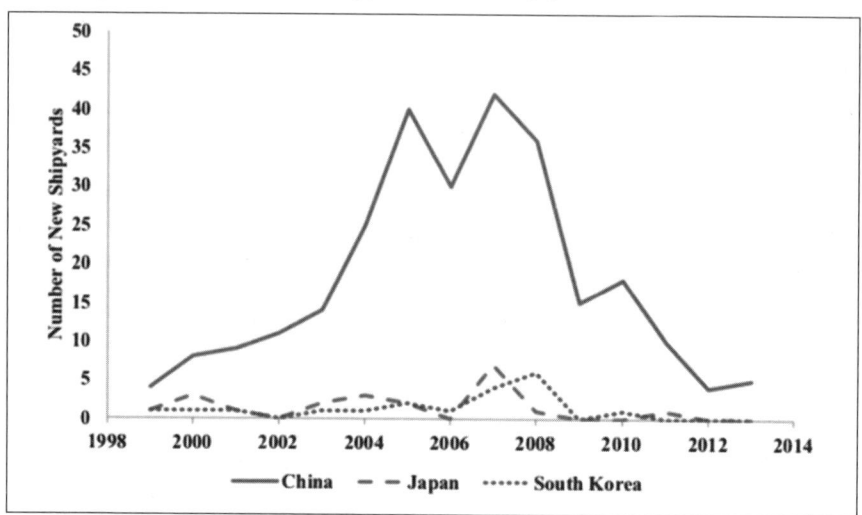

As a part of the twelfth five-year plan of 2011-15, there was a thrust towards industries producing marine ancillary equipment and many marine equipment industry clusters were formed in Shanghai, Jiangsu, Liaoning, Shandong, Zhejiang, and other coastal provinces and cities. The CSSC and CSIC got loans worth US $30 billion as part of the government's focus towards developing strategic industries. Initiatives to indigenise main propulsion have not been very successful and China continues to source them through imports. It relies on imported or license-produced diesel engines, mostly supplied by the French SEMT-Pielstick (PA and PC series), MTU Friedrichshafen GmbH, the German Shaanxi diesel engines, and Siemens AG. Gas turbines are sourced from Ukraine. Only the nuclear propulsion in the Chinese submarines is completely of Chinese origin.[22]

The Warship-Building Industry After World War II

The increased interaction of the Chinese shipbuilding industry with foreign shipbuilders helped improve the quality and efficiency of its research and development techniques, production processes, and management practices. The Chinese shipyards could gradually improve their shipbuilding capability and expand their capacity. These trends were also reflected in improvements in Chinese warships commissioned in the late 1990s and thereafter. The newer vessels are more durable, more capable of surviving damage, have longer ranges, are stealthier, and are capable of carrying a variety of modern weapon systems. China's serial production of a variety of new naval platforms, as well as the simultaneous production of several classes, stands out. China has been able to absorb some key submarine technologies from the *Kilo*-class vessels procured from Russia. This expertise was revealed in the production of new *Yuan*-class submarines which are considered to possess attributes of both the *Song-* and *Kilo*-class submarines, thus indicating the understanding of the technology by the Chinese.[23]

Table 7.1: Major Chinese Shipyards Involved in Naval Construction[24]

Ser	Shipyard	Affiliation	Warship class
1	Jiangnan-Qiuxin Shipyard (Shanghai)	CSSC	*Luyang* I and II destroyers
2	Hudong-Zhonghua Shipyard (Shanghai)	CSSC	054 class frigate, *Jiangwei* and *Jianghu* upgrades; auxiliary vessels; Type 072-III amphibious landing vessels
3	Xijiang Shipyard (Guangxi)	CSSC	Fast attack crafts
4	Huangpu Shipyard (Guangdong)	CSSC	054-class frigate; fast attack craft; replenishment vessels
5	Guangzhou Shipyard (Guangdong)	CSSC	Replenishment vessels
6	Huludao Shipyard (Liaoning)	CSIC	Nuclear submarines: 093 and 094 classes
7	Wuchang Shipyard (Wuhan/Hubei)	CSIC	Conventional submarines existing and new ones
8	Dalian Old Shipyard (Liaoning)	CSIC	*Luhai* and *Luda* destroyer class upgrade; 072-III amphibious landing vessels

Modernisation of the PLAN resulted in increased orders of warships, thereby facilitating the growth of Chinese shipbuilding. China's military modernisation is amid three interlinked transitions: land to sea, regional to global projection, and 'informatised' to 'intelligentised' warfare.[25] The transition of focus from land to sea has been emphasised on numerous forums – including the Chinese Communist Party (CCP) Politburo Meeting in 2013, when the Chinese President and General Secretary Xi Jinping stated that China needed to "do more to take interest in the sea, understand the sea, and strategically manage the sea, and continually do more to promote China's efforts to become a maritime power". President Xi also highlighted the importance of the maritime domain in the inaugural meeting of the Central Commission of Integrated Civilian and Military Development in June 2017.

China's 2015 Military Strategy White Paper also emphasised the shift of balance, stating that "the traditional mentality that land outweighs sea must be abandoned, and great importance has to be attached to managing the seas and oceans and protecting maritime rights and interests."[26]

Recent Developments in Shipbuilding

In 2015, the Chinese government announced the 'Made in China 2025' initiative strategy for Industry 4.0 like other countries – specifically, the German initiative for creating Industry 4.0. 'Made in China 2025' highlights building a smart demo ship as a priority development. Shipbuilding 4.0 in the Chinese shipbuilding industry is called 5S – a ship's operation intelligent service system that features Sea, Ship, System, Smart, and Services. The smart ship will have features including the ship safety assessment, ship energy efficiency monitoring, analysis, assessment and optimisation, status, assessment and maintenance optimisation, sea route, ship navigation and operation control, all connected via Big Data.[27]

The 'Made in China 2025' initiative aims to use the new era of smart manufacturing to make China the manufacturing superpower. The plan emphasises the importance of developing marine engineering equipment and high-tech ships. The plan for the marine sector outlines specific targets to increase local content for offshore engineering equipment and high-tech ships, expand the global market share in independent design and construction, develop standards in design, and establish Chinese companies as leading global technology providers.[28] In the thirteenth five-year plan (2016-20), the Chinese government assiduously pushed the state-run and dominated industrial base to overcome lingering technological shortcomings.[29]

In December 2017, during the Marintec China 2017 in Shanghai, a fuel-efficient bulk carrier – 38,800 DWT 'Great Intelligence' – was revealed as the first smart ship made in China. The ship was designed by the Shanghai Merchant Ship Design and Research Institute (SDARI) and built at Guangzhou Wenchong Shipyard Co. (GWS), a subsidiary of the CSSC. Llyod's Register (LR), System Engineering Research Institute (SERI), and China Class Society (CCS) were also involved in the project. The ship has features of an intelligent navigation system to optimise shipping routes and reach destinations in the quickest possible times, with minimum fuel consumption, and has features of self-learning to spot dangers and locate system bugs.[30]

As China established itself as a leader in shipbuilding, it also built its industrial capacity and achieved economic development by converting shipbuilding enterprises into 'systemic industrial organisations' that focussed on various types of ships and other related fields, such as producing equipment and materials, academic research, and education. Newbuilding statistics from IHS Markit show that China had the busiest order book in 2017, with 41 per cent of newbuilding tonnage; it remains the largest shipbuilding nation in the world.[31] China dominates the global shipbuilding market with 70 per cent of ships built being exported to 91 countries and regions, including countries like Greece, Norway, the US, the UK, Japan, South Korea, and Germany. Chinese shipbuilding output increased by nearly 13 times between 2002 and 2012, and China became the world's largest shipbuilder in 2010, far ahead of the targeted date of 2015.[32] The CSSC and CSIC are now being merged into a single entity and restructured, giving them a new scale to match Hyundai Heavy Industries' proposed acquisition of South Korea's Daewoo Shipbuilding and Marine Engineering. The mega-merger will ensure optimal capacity utilisation, reduce debt, and increase economies of scale.[33]

China has parlayed the world's second-largest economy and the second-largest defence budget into the world's largest ongoing comprehensive naval build-up. In this venture, China utilises its shipbuilding infrastructure which is now the largest in the world. Commercial production is price-capped in part by China's relatively stable business and vendor base. It helps subsidise military production. Chinese shipbuilding is greatly facilitated by an organisational structure for collecting and disseminating technology and integrating it into development and production processes at an industrial scale. High-tech, high-value-added, and high-reliability commercial shipbuilding – for example, of liquid natural gas (LNG) and liquid propane gas (LPG) tankers, very large crude carriers (VLCCs), high-capacity container ships carrying more than 10,000 twenty-foot equivalent units (TEUs), and even cruise ships – can be directly relevant to warship production.[34]

NOTES

1   Nitin Agarwala and Rana Divyank Chaudhary, "Growth of Shipbuilding in China: The Science, Technology, and Innovation Route," (Occasional Paper no. 31, Institute of Chinese Studies, Delhi, May 2019), 29.

2   Tiezzi, "Chinese Naval Shipbuilding: Measuring the Waves".

3   Peter Lorge, *War, Politics and Society in Early Modern China, 900-1795* (New York: Taylor & Francis, 2006), 1–4.

4   Peter Lorge, "Water Forces and Naval Operations," *A Military History of China*, ed. David Graff and Robin Higham (Boulder, CO: Westview Press, 2002), 81.

5   Lorge, "Water Forces and Naval Operations", 86.

6   Edward L. Dreyer, *Zheng He: China and the Oceans in the Early Ming Dynasty, 1405-1433,* (New York: Pearson, 2006), 50–70.

7   Chaudhuri, *Trade and Civilisation in the Indian Ocean,* ch.2, 15.

8   Maddison, *The World Economy: A Millennial Perspective,* ch.2, 117–18.

9   TG Moore, *China in the World Market: Chinese Industry and International Sources of Reform in the Post-Mao Era,* (Cambridge, UK: Cambridge University Press, 2002), 169.

10 Maddison, *The World Economy: A Millennial Perspective,* ch.2, 146.

11 Moore, *China in the World Market,* ch.7, 170-72.

12 Moore, *China in the World Market,* 175–77.

13 Agarwala and Chaudhary, "Growth of Shipbuilding in China", 6.

14 Geoffrey Murray, "China Charts Course into LNG Shipbuilding - Global Times," *Global Times,* http://www.globaltimes.cn/content/841637.shtml, accessed 30 June 2019.

15 Tian C, "The Six Decades of Chinese Industry," *The Economic Observer,* http://www.eeo.com.cn/ens/news/, accessed 02 July 2019.

16 Russell Smyth, Xin Deng, and Junli Wang, "Restructuring State-Owned Big Business in Former Planned Economies: The Case of China's Shipbuilding Industry," *New Zealand Journal of Asian Studies* vol. 6, no. 1 (June 2004), 30.

17 Zhong Nán, "Building a New Marine Economy," *China Daily,* http://www.chinadaily.com.cn/kindle/2017-07/23/content_30215621.htm, accessed 30 June 2019.

18 Moore, *China in the World Market,* ch.7, 164-65.

19 Myrto Kalouptsidi, "China's Shipbuilding Industry: Measuring the Effect of Industrial Policy," *LSE Business Review,* https://blogs.lse.ac.uk/businessreview/2019/04/15/chinas-shipbuilding-industry-measuring-the-effect-of-industrial-policy/, accessed 15 April 2019.

20 Kalouptsidi, "China's Shipbuilding Industry".

21 Kalouptsidi, "China's Shipbuilding Industry".

22 Sarah Kirchberger, *Assessing China's Naval Power: Technological Innovation, Economic Constraints, and Strategic Implications* (New York: Springer, 2016), 143–45.

23 ES Medeiros et al, "China's Shipbuilding Industry," *A New Direction for China's Defense Industry* (Santa Monica, CA: Rand Corporation, 2005), 148.

24 Medeiros et al, "China's Shipbuilding Industry", 124.

25 Tate Nurkin et al, "China's Advanced Weapons Systems", *US-China Economic Review Commission,* https://www.uscc.gov/research/chinas-advanced-weapons-systems, accessed 24 June 2019.

26   Nurkin et al, "China's Advanced Weapons Systems", 28.

27   Venesa Stanic et al, "Toward Shipbuilding 4.0-an Industry 4.0 Changing the Face of the Shipbuilding Industry," *Brodogradnja* (September 2018), 69.

28   Liwei Tan, "Made in China 2025 Policy: Maritime Equipment and High-Tech Ships" *Maritime Business Day,* https://www.slideshare.net/FinproRy/made-in-china-2025-policy-maritime-equipment-and-hightech-ships, accessed 30 June 2019.

29   Jon Grevatt, "A Great Leap Forward," *Jane's Defence Weekly,* http://janes.ihs.com.inelibrary.remotexs.in/DefenceWeekly/Display/jdw65559-jdw-2017, accessed 30 June 2019.

30   "China's 1st Smart Ship Makes a Debut," *World Maritime News,* https://worldmaritimenews.com/archives/237342/chinas-1st-smart-ship-makes-a-debut/, 26 June 2019.

31   Nick Savvides, "Outlook 2018: Asian Shipyards to Embrace Innovation in 2018," *IHS Markit Safety at Sea,* https://safetyatsea.net/news/2017/outlook-2018-asian-shipyards-to-embrace-innovation-in-2018/, accessed 07 January 2019.

32   Andrew S Erickson, ed., *Chinese Naval Shipbuilding: An Ambitious and Uncertain Course* (Annapolis, Maryland: Naval Institute Press, 2017), 7.

33   "CSSC-CSIC Megamerger Confirmed at Last," *The Maritime Executive,* https://maritime-executive.com/article/cssc-csic-shipbuilding-megamerger-confirmed-at-last, accessed 03 July 2019.

34   Andrew S Erickson, "Chinese Naval Shipbuilding: Full Steam Ahead," *The Maritime Executive,* https://www.maritime-executive.com/editorials/chinese-naval-shipbuilding-full-steam-ahead, accessed 18 January 2019.

8

Case Study – India

Introduction

The fourth and final case study is on India, without which no analysis of the Indo-Pacific can be considered complete. India has a population of 1.3 billion, which makes it home to one-fifth of the world's population, and is the largest democracy. It has an area of 3.287 million square kilometres and a coastline of 7,516.6 kilometres. India is mountain-guarded in the north and sea-grit in the south, so its geography provides it with natural isolation. However, it has a history of intense interaction and exchange with virtually every major civilisation. The physical isolation of geography facilitated the independent evolution of Indian civilisation, while the intense interaction with varied global cultures contributed to its diversity. As a result, India evolved into a culture with unity through diversity.[1] Although the origins of Indian culture date back to the Indus Valley Civilisation that existed around 3000 BC, it was between 1300 to 400 BC – during the Vedic era – that settlements emerged as a network of small kingdoms and republics, many of them centred around towns; that was the first time there was an awareness of the whole subcontinent as a geographical and civilisational unit.[2]

There are some aspects of modern Indian society that provide evidence of remarkable social-cultural continuity over the ages, showing that some of the civilisational traits have survived over millennia. For example, the oxcarts of the Harappan civilisation can still be seen in many parts of rural India; the Gayatri mantra, a hymn contained in the Rig Veda, composed four millennia ago, is still chanted by millions of Indians; the Sentinelese tribe of the Andaman

Islands deliberately retains its stone age culture; and some of the placenames from the Vedic ages are still in use in many parts of the country.[3] Two great epics from the Vedic period, the Ramayana and Mahabharata, depict a lot about how the geographical conception of India evolved in the Iron Age and continue to be central to its culture and daily life. An interesting detail of these epics is their cardinal orientation along the major trade routes. The geography of Ramayana is oriented along a North-South axis,[4] while the Mahabharata is generally oriented on an East-West axis.[5] Both these routes facilitated the trade of goods, as also the flow of intellectual philosophies of the Upanishads, Mahavira, and Gautam Buddha.[6] The Uttara Path was a well-trodden route by the Iron Age that was also formalised during the Mauryan Empire. Since then, it has almost continuously been rebuilt in some approximation to the original. Sher Shah Suri, the Mughals, and the British invested heavily in maintaining it.[7]

These trade routes connected the hinterland to major port cities. Whilst the west coast traded with West Asia and the Graeco-Roman world, the east coast of India traded with Southeast Asia, extending all the way to China. Indian civilisation exerted influence on Southeast Asia almost exclusively through trade with no record of Indian military intervention in the region – the exception remains the Chola raids on Srivijaya in the eleventh century. The cultural impact and civilisational influence still live on in the entire region, which includes Myanmar, Malaysia, Thailand, Indonesia, and the Korean peninsula.[8]

The flourishing maritime trade and significant cultural influence of India in the entire Indian Ocean Region (IOR) was made possible because of the advanced shipbuilding industry which existed since early history and evolved with time. Indian shipbuilding faced a setback under colonial rule when it missed the Industrial Revolution, with its growth being curtailed to meet the requirements of the British. After gaining independence in 1947, the Indian Government has taken many initiatives to revive the shipbuilding sector; presently, the Indian shipbuilding industry is the fourteenth largest in the world, while the Indian Navy is the fifth largest navy with around 200 ships.

History Before the Mid-Nineteenth Century

The shipbuilding prowess of India can be traced back to the Indus Valley Civilisation,[9] when the Harappan city of Lothal had one of the oldest tide docks of the world, built around 2300 BC. Lothal developed as a large emporium and servicing station at the head of the Gulf of Cambay, in the estuary of the Sabarmati and Bhogava rivers.[10] The entire coastline of present-day Kutch, Kathiawar, and south Gujarat was studded with Harappan ports which were used for extensive trade with Mesopotamia. The Harappans used flat-bottomed boats without sails in rivers and creeks; and ships with sails, high prow, and sharp keel on high seas. These sea-going vessels could be berthed in the deep waters of the gulf.[11]

During the Vedic period,[12] Indians had an advanced shipbuilding industry which produced large ships that were used extensively for international interaction and maritime commerce. Sanskrit literature in all its forms such as the Vedas, Sutras, and Puranas, as well as Buddhist literature like the Chronicles of Ceylon, Jatakas, etc., have references to India's maritime trade.[13] In the Mauryan Period (600 to 200 BCE), there was high demand for both river and ocean traffic, stimulating a flourishing shipbuilding industry that provided employment to many. The Arthashastra[14] contains elaborate descriptions of the shipbuilding and maritime activities of the time. The shipbuilding industry was state-controlled and a monopoly of the government. Shipbuilders were a class of artisans who were salaried public servants. These ships were built in royal shipyards and let out on hire, both to those who undertook voyages and to professional merchants. The war office of Emperor Chandragupta (321 to 297 BCE) had a Board of Admiralty and a Naval department which oversaw national shipping. This has also been described in detail in the Arthashastra.[15] The Naval department was headed by the *Nawdhyaksh* or Superintendent of ships. The naval organisation and shipbuilding activities were similar even during the subsequent periods of Ashoka the Great.[16] The Arthashastra has passages that compare the land and sea routes, indicating that goods were transported using the coastal as well as mid-sea routes to destinations in the Red Sea or Southeast Asia.[17]

The Age of the Mauryans – of Chandragupta and Ashoka, was followed by the Age of the Andhras of the South, and the Kushanas of the North, which continued the development of foreign trade and international interaction. During the Andhra Period – in around 75 AD, seafarers from Chinga (Kalinga) undertook voyages into the deep seas in the east, reaching the Indonesian archipelago and establishing trade links with Java.[18] The

Satavahana Dynasty (200 BCE to 250 AD) maintained a massive fleet of sailing ships and there was trade – both over land and by sea with western Asia, Greece, Rome, and Egypt, as well as with China and eastern Asia.

As the demand for trade grew from Rome, in about 47 AD, a pilot named Hippalus discovered the regularity of monsoons in the Indian Ocean, and identified the season when the annual winds settle in the north and blow continually from the south-west.[19] After this discovery, ships began to sail directly from Rome to India's Malabar Coast without hugging the coast, and were hence safe from Arab pirate attacks.[20] The multi-oared galleys gave way to sail ships, and oars were retained only for manoeuvring in the harbour and for use when winds dropped at sea. The subsequent empires of the Indian subcontinent continued to extend maritime trade and influence from Europe to the Far East and this was supported – in fair measure – by an advanced and innovative shipbuilding industry. During the Gupta Empire (fourth to sixth century CE), astronomers like Aryabhatta and Varahamihira formulated a method of fixing a ship's position with respect to the stars. During the Chola Dynasty (ninth to eleventh century CE), clay oil lamps were innovatively placed atop palm trees at ports, serving the purpose of lighthouses by facilitating navigation by night. The ships were built of fir-timber and were double-planked.

King Bhoja Narapati who reigned in the eleventh century CE in Dhara (modern Dhar of central India) is believed to have authored the Yukti Kalpa Taru, a treatise on the art of shipbuilding in ancient India.[21] The book contains an elaborate classification of ships – based on their sizes and roles – into two primary classes, 'Samanya' designed for river traffic and waterways, and 'Visesha' for sea-going vessels. Each of these is further classified into many sub-categories with details of design and construction guidelines. It specifies the kind of wood to be used for various classes; procedures for joining planks; directions for decorating and furnishing ships; paint schemes for masts, cabins, holds, etc.; and includes details of the cabins and specifics for designing ships for naval warfare. One of the classes called 'Agramandira' recommends a design with cabins towards its prows, suitable for long voyages and naval warfare. A remarkable indication of civilisational continuity is depicted by the fact that many of the specifications match with the ships which are referred to in the Ramayana and Mahabharata.[22]

During the Hindu Period (1175 to 1572 AD),[23] successive rulers who reigned in different parts of the Indian subcontinent continued to support advances in the shipbuilding industry and dominated maritime trade in the region. The Pandyas had excellent knowledge of metallurgy and were adept at working with copper alloys, such as brass and bronze. The Vijayanagara Kingdom had around three hundred ports, an efficient port organisation that managed international trade, and built ships in many centres in India and the Maldives.[24] Whilst the ships built in India were not as large as the Chinese Junks, they were larger than those built by the Europeans. All through the medieval period (1378 to 1797 AD),[25] the ships built in India were compact and had high manoeuvrability in shallow waters, ideal for small ports and estuaries in India and eased navigation in the Persian Gulf and the Red Sea. During this period – for the first time – Indians began to construct ships purely for war at sea.

From the early days of the Mauryan Empire, dynasty after dynasty expanded international maritime trade and this trend continued well up to the medieval period, until the arrival of foreign powers on Indian soil. This remarkable continuity carried forward India's maritime civilisational legacy which was firmly supported by an advanced indigenous shipbuilding industry and monopoly of certain Indian products in the global market. The Indian products that dominated the markets – extending from the West to the East — included renowned art industrial fabrics; manufactured luxuries; end-products of advanced applied chemistry, like dyes;[26] cosmetics; artificial imitation of natural flower scents; 'Vajralepa' (a cement-like material touted to be as strong as thunderbolt);[27] and tempered steel that was made by applying methods of advanced metallurgy. In exchange, India received large quantities of gold and precious metals, boosting its economy and furthering its stature as a world power. Trade with the Far East extended India's influence up to Java, Sumatra, and Cambodia during the Gupta Period (320 to 550 AD), which was further extended to China and Japan during the Reign of Harshavardhana (606 to 647 AD) and Pulakesi of the Chalukyas (sixth to twelfth century), and the Cholas (ninth to thirteenth century).[28]

A notable feature of Indian shipbuilding was that the shipwrights were mostly Hindus, and their profession was transferred from father to son. Under the prevalent caste system, the division of labour was such that the skilled workers in shipbuilding were of a "low caste", and since the learning of science was largely the preserve of the higher castes, advancements in shipbuilding came from on-job innovation and imitation from international ships.[29] However, the quality of workmanship was world-class, and Indian-made vessels were recognised for their finish and durability. In seventeenth-century Europe, as a part of the Scientific Revolution which preceded the Industrial Revolution, science was separated from religion and was integrated with technology, so that advancements in technology now originated from scientific research. As this did not happen in India – and more specifically in Indian shipbuilding, the consequences started becoming apparent in subsequent times, in terms of the introduction of new technologies in ships. However, some exceptions stood out, like the reputed Parsi shipbuilders of Surat who subsequently moved to Bombay in the early eighteenth century. Lowjee Nasserajee and his successors — who earned the title "Master Builders" – constructed many warships of repute for the Bombay Marine and later, for the Royal Navy.[30]

Starting from the sixteenth century, there were successive conflicts and competitions for trade and dominance by external powers as well as indigenous resistance movements. Some of the major powers were the Mughals (1526 to 1858), the Marathas (1674 to 1818), the Portuguese (1505 to 1961), the Dutch (1605 to 1825), and the British (1612 to 1947). The Portuguese – who were the earliest of the European powers to arrive in Goa in 1498 – ushered an era of warships with guns mounted on board, and they set up shipbuilding yards in Goa and Daman, employing Indian artificers and using indigenously sourced material. The Indians were quick to emulate, and the Zamorins of Calicut were the earliest to start building warships with similar guns on board. The Marathas were, however, the most versatile, and hired Portuguese specialists as a strategy to absorb new technologies and build ships in their yards in Vijaydurg, Swarndurg, and Kolaba. Shivaji also hired Portuguese like Rui Leitao Viegas in his navy as mercenaries, effectively achieving seamless integration between the technology capability and concept of operation of his naval forces.[31]

By the early eighteenth century, the Maratha navy under Admiral Kanhoji Angre had a massive armada that dominated the entire coast, from Savantwadi to Bombay. He had dockyard facilities for building vessels, mounting guns and preparing them for sea.[32] The Maratha shipbuilding centres extended all the way from Konkan, Bombay to Surat.[33] During the declining days of the Mughal Period around the late eighteenth century, Parsi shipbuilders established a legacy of the Wadia shipbuilding lineage.[34] The Duncan dock at Bombay was constructed in 1810, and by 1818, a seventy-four Guns Teak built-ship the *Malabar* was launched. Subsequently, a series of ships of much higher capabilities were built by the Bombay dockyard under the renowned builder Jamsetjee Bomanjee.[35]

The British, however, had the most profound impact on Indian shipbuilding. Interaction with the British started towards the end of the sixteenth century, when they captured two Portuguese ships carrying merchandise from India. On inspection of cargo and the transit papers, the British realised the enormity of trade potential in India and the actual purchase price of spices. The Portuguese and Dutch had been selling calicoes, silks, gold, pearls, drugs, porcelain, ebony, etc., in Europe at very high prices. Because of the revelations, the Queen of England issued a Charter in 1600 to set up the British East India Company to establish direct trade with India. By 1700, the Company's trade had grown so much that it managed to establish itself as a power through territory acquisition, taking advantage of political disorders in India.

The British encashed on their maritime supremacy to gradually monopolise the foreign trade of India and consequently effect a fundamental change in the mode of trade. Until then, many Indian merchants had their own ships and were often the shippers of their own products. The British began to enforce trading using their own ships, and gradually, the Indian merchants were reduced to just being buyers and sellers – confined to their shores. The British began more focussed consolidation of their control over India after the loss of the American colonies in 1776, and their victory over Napoleon in 1815. Sea trade was subsequently organised as a Joint Stock Company through a Royal Charter, and the Indian Empire was transferred to the British Crown.[36]

From the early nineteenth century, the British began to use their monopoly in trade to make it difficult for Indian shipping companies to operate. By an Act of 1814, a ship trading with England was likely to be forfeited unless 75 per cent of its crew and the captain were British. The British Government tariff policy of 1812 mandated a 15 per cent import duty on goods being brought in Indian ships as against the 7.5 per cent in the case of British ships.

The Indian shipping companies could not even compete in coastal trade since British companies were given contracts for government goods. Mahatma Gandhi once stated, "Indian Shipping had to perish so that British Shipping might flourish."[37]

Mid-Nineteenth Century to the End of World War II

From the early nineteenth century, inventions of the Industrial Revolution began to majorly impact ship designs. Paddlewheel steamers were invented in 1807, while screw propellers were invented in 1836. Some ships like the *Assaye* and the *Punjaub* – which were built as sailing ships in 1854 and 1856 in India – were later converted to paddle wheel steamers, and then, to screw-propelled ships in Europe. Iron hull was introduced in 1840, and steel hull in 1880. Prime movers were also introduced in quick succession. The diesel engine was introduced in 1895; in 1900, the turbine or rotary engine; and in 1903, the electric motor vessel was introduced. The fuel used also changed from wood, coal, oil, and finally, to gas – in that sequence. The complexity of ships led to the transition from the single designer concept to design teams.[38] Whilst these technology transformations completely changed the world of shipbuilding, Indian companies lost out. Therefore, the advent of steel steamships and advancements in prime movers, combined with a structured strategy of the British to kill the Indian shipbuilding industry, resulted in the latter's ultimate demise.[39]

Towards the end of the nineteenth century and early twentieth century, many Indian entrepreneurs ventured into Indian shipping and started their own shipping companies, despite resistance from the government and prohibiting policies. Some of the examples include the Tata Line in 1894, the Swadeshi Line in 1906, the Bengal Steamship Company in 1905, and the Mogul Line in 1877. However, most of them went into liquidation since the established companies resorted to unfair trade practices, protected by the British.[40]

On 27 March 1919, Sheth Narottam Morarjee and Shri Walchand Hirachand set up the Scindia Steam Navigation Co. Ltd., purchasing a ship *Loyalty* from Maharaja Madhavrao Scindia. The *Loyalty* sailed from Bombay to Great Britain, unfurling the flag of the Indian merchant marine[41] on 5 April 1919.[42] The British made the Scindia Company sign agreements in

1923, and later in 1933, prohibiting them from international trade and confining them to coastal trade only. Whilst these agreements continued to be in force till 1950, after independence in 1947, the Government of India permitted the Scindia Navigation Company to use *Liberty*-class ships that they had procured from the US to ply in the routes, from India to the US and Canada. Parallely, right from its inception, Narottam Morarjee and Walchand Hirachand also pursued efforts to revive indigenous shipbuilding and explored the feasibility of setting up a modern shipbuilding yard in India, in consultation with shipbuilding experts in Great Britain.

The colonial British government did not favour the setting up of such a strategic industry under the ownership of Indian entrepreneurs, therefore, sparing no expense to discourage the initiative. Despite all the hindrances by the then-ruling government, the Great Depression of the 1930s, and the World Wars, the foundation stone of the shipyard was laid in Visakhapatnam in 1941, and the construction of the shipyard was completed in 1946. The consultants under Sir Alexander Gibb and Partners recommended Visakhapatnam as an ideal site for a modern shipyard, and designed it to be a shipyard with eight building berths and requisite workshops. Visakhapatnam was a natural land-locked harbour on the Coromandel Coast, midway between Calcutta and Madras. Well-connected to Tatanagar,[43] from where steel was available – a critical raw material required for shipbuilding, Visakhapatnam had proximity to a major fresh water source, adequate depth alongside, and a ready availability of unskilled labour force. Although it was the first modern shipyard in India, it faced very unfavourable market conditions since it had no subsidies from the government – unlike the British, Italian, and Australian shipbuilders who not only got subsidies in material sourcing and taxation rebates, but also had assured orders from their shipping companies who were mandated to use indigenous bottoms for coastal trade under the Cabotage Law.[44]

After World War II

During World War II and during the Bengal Famine of 1943, the Government of India under the British felt the shortage of Indian-owned ships and inadequacy of Indian shipping, and therefore set up a Reconstruction Policy Sub-Committee on Shipping in October 1945. The mandate of the sub-committee was to recommend corrective measures to resurrect the Indian shipbuilding industry, not only for commercial reasons but also because of the development of the Royal Indian Navy. The Policy Sub-Committee's report was submitted and approved by the Government in March 1947, and the decisions were published as a Resolution on 12 July 1947. It recommended that the entirety of coastal shipping and half the international trade should be carried out using Indian-built ships. After independence, the Government of India announced its industrial policy on 6 April 1948 in which it adopted a mixed economy policy. Shipbuilding was thus made open to private entrepreneurs although the state was to participate in it.

The Scindia Shipyard in Visakhapatnam – the first modern shipyard of India – commenced production almost as soon as two building berths were completed, laying the keel of the first ocean-going ship of 8000 tons *Jala-Usha* in June 1946, and building four more ships of the same type up to 1948. These ships were costlier than British ships of the same size (68 lakhs against 44 lakhs) but were still less costly compared to those built in Australia and Italy. The production cost of a ship built in Scindia Shipyard was higher than contemporary shipyards in the West because the initial cost of setting up the yard was considerably higher, due to many reasons. The disruptions in construction and supplies during World War II had added to the initial investment required to set up the yard. The cost of indigenous steel delivered to Visakhapatnam was considerably higher than the cost at which steel was available to British shipyards. Most of the propulsion and auxiliary machinery, instrumentation, and stores had to be imported, and the locally available manpower was unskilled and inexperienced, while the manpower hired from abroad was expensive. Finally, the shipyards in Great Britain were of much higher capacity and had been in production for many years, which was why the overheads were considerably lower.

Despite the higher cost of production, the shipyard continued to build ships for its own shipping company and built four 8000-ton ships till 1948. However, the model was not sustainable; therefore, the shipyard approached the government in 1948 to provide subsidies to enable Indian shipping companies to purchase ships built in India, or alternately take over the shipyard. The government did not take over the shipyard due to disagreements in the evaluation of the shipyard cost, but agreed to place orders for a few ships until the issue was resolved. Subsequently, based on the recommendations of the Estimates Committee, the government took over the shipyard in 1952. Hindustan Shipyard Limited was established, with two-thirds ownership by the government and one-third by the Scindias.

In July 1952, an agreement was signed with a French shipbuilder – Ateliers et Chantiers de la Loire (ACL) – to provide technical aid in the management-and operation-of the yard. The charter included training Indians in design, production, and marketing. The French were also to help in securing international orders and establishing logistic chains for the supply of essential input material. However, despite the consultation and government takeover, the shipyard fell short of the targets set in the first- and second five-year plans. Though there were technological advancements made by the shipyard – which graduated from building only steamships to building diesel engine-fitted Maierform ships – the overall yearly gross tonnage constructed and timely delivery targets were missed. The contract with ACL was eventually terminated in 1958, but the shipyard built a new series of ships of 9500 to 12000 DWT which could achieve 17 knots speed, based on the designs from West German shipbuilder M/s Flender-Werke (located in Lübeck).

Through the period of the first two five-year plans, the Estimates Committee made a series of recommendations, from time to time, to reduce the cost of ship construction and achieve timely completions. The committee recommended setting up planning and design departments in HSL, creating training facilities to produce skilled local workforce, making standard designs for shipbuilding to cut costs, etc. However, during this period, the government continued to give fixed-price contracts and expand the facilities in the yard in a phased manner. All through the decade, the government provided a subsidy ranging from 25 to 35 per cent to HSL.

In 1956, the government took two major initiatives to enhance the shipbuilding capability of the country. One, it approved three standard ship designs which could be used by Indian shipyards. These included a 9500 tonner with speeds of 16 to 17 knots for overseas trade, an 8000 tonner with speeds of about 12 knots, and a 5000 tonner with speeds of about 12 knots – the latter two for coastal trade. Subsequently, the Estimates Committee also added one more design of 2500 to 4000 tonner ships for traffic in smaller ports. The machinery and auxiliary requirements for all these designs were finalised and standardised. The second major initiative of the government was the decision to establish a factory for manufacturing propelling machinery in the public sector. Towards this, an inter-ministerial committee was set up to seek proposals for licence production from major global manufacturers.

Towards the end of the first five-year plan, the Government of India decided to set up a second major shipyard in the country, with foreign technical collaboration. Under the Colombo Plan,[45] the UK government offered services of an expert committee, under the leadership of Mr James Lenaghan in November 1956. Based on the recommendations of the expert committee, the government decided to set up the second major shipyard in Cochin (Ernakulam) during the third five-year plan. The shipyard was to have the capacity to build 60,000 GRT, expandable to 80000 GRT, and build 9000 to 12000 tonners as well as Naval ships. The Cochin Shipyard was finally conceived in 1969, and was designed and constructed under the technical collaboration with M/s Mitsubishi Heavy Industries (MHI), Japan and commenced shipbuilding in 1978.[46]

During the third and fourth five-year plan periods (1961 to 1974), the GoI primarily focussed on the expansion of HSL and the setting up of CSL. There were also plans of setting up marine diesel engine assembly facilities under the licensed production of a major global manufacturer, under the aegis of a public sector enterprise. The manufacture of smaller vessels for coastal traffic, boats, and tugs was to be handled by the private sector. During the fifth plan period (1974 to 1979) – whilst HSL and CSL were still producing below their capacity – the GoI was also considering setting up one more shipyard. However, in the following decade, by the end of the seventh five-year plan in 1990, global demand for shipbuilding was low and therefore, the GoI shelved the idea of setting up a third major shipyard. CSL and HSL continued to produce below capacity and MDL and GRSE had failed to get any commercial shipbuilding orders. The GoI decided to set up a nodal design agency, initially in HSL, to provide services to all the shipyards in the country and the National Ship Design and Research Centre (NSDRC) was set up in 1993. During the ninth five-year plan period (1997-2002), the shipbuilding sector was delicensed except for warship building, and considering the low-capacity utilisation of the public sector shipyards, the GoI was looking for disinvestments.[47]

In the tenth plan period (2002-2007), India's share of global shipbuilding increased from 0.1 per cent to 1.3 per cent in 2006. Public shipyard CSL and private shipyards ABG, Bharati, and Chowgule managed export orders and many of the shipyards had full order books by 2007. Private shipyards were constrained by their limited capacity to take more international orders, but there was high profitability in international orders of ship repairs. In the following years, there was a slump in the shipping trade, and consequently, the shipyards once again faced a slowdown. The twelfth plan set an ambitious target – achieving 5 per cent of global shipbuilding and 10 per cent of global ship repairs. Considering that 95 per cent of India's foreign trade by volume and 65 per cent by value is through sea, the GoI has consistently been working on policy reforms to increase the percentage of sea trade being carried by Indian bottoms. In the 1980s, about 60 per cent of EXIM trade was serviced by foreign-flagged vessels, and in 2009-2010, this increased to about 92 per cent – this indicates that as the economy, and hence, trade grew, shipbuilding lagged further, and a larger share of trade was handled by foreign ships.[48] The two-fold impact of such a trend is that more foreign exchange is drained and the profits from shipping do not get roped back to improving the very shipbuilding industry. The NITI Aayog, in its 'Strategy for New India @ 75', has recommended that a scheme to finance shipbuilding – which is like the subsidy that was in place from 2002 to 2007 – should be introduced to boost indigenous shipbuilding.[49]

The Warship-Building Industry After World War II

In 1947 when India gained independence, there was no shipyard in the country that could build warships. The last time a warship had been built in India was in 1848, when an 80-gun wooden warship *Madras* was built by the Mazagon dock. However, the Peninsular & Orient Company (P&O) had continued to maintain two of its small shipyards – Mazagon Dock in Bombay and Garden Reach Workshop (GRW) in Calcutta – primarily to maintain ships owned by the company. In 1956, the P&O Group offered to sell these shipyards to the Government of India (GoI). However, a GoI takeover was not recommended by the Planning Commission on the grounds that there was not much demand for vessels of 4000 GRT and yard crafts that these yards could build – albeit with some initial investment to upgrade the infrastructure. In 1957,

Mr VK Krishna Menon – the then Defence Minister – steered a package deal on behalf of the Ministry of Defence to purchase the two yards for about Rs 4 Crores, in order to facilitate the construction of warships for the Indian Navy.

Mazagon Dock Limited (MDL) was thus established on a 35-acre site. Subsequently, the Kassara basin to its north – under the Bombay Port Trust, and the Mahindra Jeep Assembly Plant, a plot to its south – used by M/s Mahindra, were taken over by MDL in 1963. M/s Alexander Gibbs and Partners who were consultants for the expansion of the Naval Dockyard Bombay, were appointed as advisors for augmenting the facilities of MDL for ship repairs and ship construction. The expansion of MDL was completed in the next five years. Simultaneously, India began to evaluate experienced shipbuilders in Europe who could provide the design, training, and apprenticeship to set up the infrastructure for warship production capability in India. Shipbuilders in Sweden, the Netherlands, and the UK's Admiralty were evaluated. Eventually, the GoI decided to work with the UK's Admiralty and their shipbuilders – M/s Yarrow & Co. Ltd. and M/s Vickers Armstrong Ltd. – from 1960 to 1964, on a proposal for building *Leander*-class frigates in India. This proposal was also facilitated by an offer of a special defence credit of 4.7 million pounds by Britain in order to cover the external cost of the frigate project, as well as the expansion of MDL.[50] The first of the *Leander*-class ships was launched in 1968, just as the expansion of MDL was complete. In the next fifteen years, 6 *Leanders* were built by MDL, delivering each ship in about 30 months – the last one INS *Vindhyagiri* was commissioned in 1981.

The decision to set up warship-building infrastructure and build warships under the transfer of technology and design from international experts helped India to put in place a structured system, which became the bedrock of an advanced indigenous warship-building capability. The venture succeeded due to the unified teamwork of the Ministry of Defence, the Indian Navy, and MDL. The Admiralty and its shipbuilders, Yarrow and Vickers, provided the design; guidance drawings; and theoretical and on-job training to personnel from the IN and the MDL on warship building and associated activities, like stores procurement and management, nuances of warship design, overseeing and acceptance trials, documentation, training, etc. As a part of an all-encompassing consultancy and apprenticeship under the overall guidance of the Admiralty, the British shipbuilders provided the complete design of the latest version of the *Leander*-class frigate, FSA34,[51] which was under construction in the UK. Indians also gained hands-on experience in warship design and the preparation of detailed production drawings. About 300 Indians were trained in the UK for varying durations, and about 60 British specialists from various fields of specialisations trained and worked in MDL.[52] Most of the equipment, including propulsion machinery, auxiliaries, weapons, and stores were initially sourced under the guidance of the British shipbuilders from the same sources that supplied to the frigates under construction in the UK, and indigenised progressively through the project. For example, the boilers were made in Naval Dockyard Bombay, the turbines were manufactured under the transfer of technology by BHEL and Radars, and the communications and electronic warfare systems from Bharat Electronics – all public sector organisations. By the time the sixth and final *Leander* was built, the indigenous content was about 70 per cent.[53]

The early impact of the Naval Design group began to show effects, in terms of design modifications to meet specific Indian requirements. The Indian Navy decided to change the weapons and sensors fitted on the second and third frigates to those being fitted on the *Leanders* being built in the Netherlands. Dutch systems like the Fire Control Systems, radars, and displays were integrated with assistance and training from the Netherlands Design Bureau – Nevesbu – under which the personnel from MDL and the Naval Design Group worked with the Dutch to prepare detailed drawings for the modified configuration. This not only gave the Indians confidence to modify the design and integrate other systems to the original *Leander* design of the British, but also introduced them to a new method of working, wherein complete detailing – up to the level of fasteners and doors – is finalised early on, at the design drawing stage. This was the practice being followed by the Dutch, Americans, and Japanese. In the fifth and sixth frigates, more design changes were made to enhance anti-submarine warfare capabilities, operation of a Sea King helicopter, and fitment of an Italian Electronic Warfare system integrated into other ship sensors.[54]

Under the Defence Plan 1964-69, the Indian Navy planned the expansion-and enhancement-of capabilities in the mid-1960s, and was looking to induct submarines, destroyers, and missile systems. The UK, or other established builders in the West, were the natural choice for collaboration because of the comfort levels due to past association. India had kept track of developments in submarines and was specifically monitoring the development of the *Daphne*-class by France, *Dolphin*-class by the Netherlands, the *Oberon*-class by Britain which had evolved from the erstwhile *Porpoise*-class, and the American conventional submarines – with enhanced underwater propulsion power. India was also looking to acquire the *Daring*-class of destroyers from the UK. Pakistan became part of SEATO[55] in 1954 and CENTO[56] in 1955, and was to be given extensive naval assistance from the US and UK. Indians were considering early upgrades of Naval capabilities, given the belligerent posturing by the Indonesians in the Nicobar Islands region, and Pakistan in the Rann of Kutch region.

The Americans refused to offer submarines, and whilst the British were ready to build one *Oberon*-class in their shipyard, they expressed their inability to agree to a soft loan and deferred payment terms, owing to their own domestic financial situation. However, in 1964, the US provided Pakistan the submarine *Ghazi* for an initial loan period of three years. Since the technologies required by India were not made available by the West, the Soviets offered ships and submarines, as also the training and setting-up of repair and maintenance infrastructure. In the following decade, India acquired several *Foxtrot*-class submarines, *Petyas* and missile boats. Indians were exposed to Soviet warships that were densely packed with equipment; followed a maintenance philosophy that was based on the running hours of exploitation; and operating, maintenance, and logistics procedures that were quite different from the West. The successful integration of the Soviet assets in the Indian Navy and the success of employing missile boats in the 1971 conflict further enthused the Soviets to offer more options to the Indians. Subsequently, more advanced *Kilo*-class submarines and *Kashin*-class destroyers followed. Gradually, the Soviet-origin weapons began to be retrofitted by Indians in the older Western-origin ships, and finally, Indians began to design ships with weapons and sensors from Soviet suppliers.[57]

Although the maintenance and deployment of warships and submarines built on a completely different design philosophy of construction and usage posed a major challenge to the Indian Navy, its homegrown expertise in warship design and interfacing weapons and sensors gave it the unique capability to build hybrid designs,[58] in collaboration with manufacturers of weapons and sensors across the globe. The hybrid designs that followed were to cater for gas turbine propulsion, installation of surface-to-surface and surface-to-air missile systems, and the main armament in the fore part of the ship. This made the ship longer and thinner in shape, while still catering for varying specifications of the power supplies and nuances of interfaces.[59] India has since added a generation of stealth frigates, destroyers, cruisers, aircraft carriers, and submarines through the decades, and has a robust ongoing warship-building programme, with state-of-the-art weapons and sensors supporting modern network-centric operation capability. In addition, India has continued with the outright purchase of warships and building submarines in collaboration with European countries, as well as undertaking joint development of weapons and sensors with the world's established manufacturers. Indian hybrid designs combine the prowess of international weapons and sensors integrated with modern ships, designed to perform optimally in the tropical conditions prevalent in the northern Indian Ocean (high temperatures, high humidity and warm, highly saline, highly corrosive seas) – uniquely suitable for export to countries in the Indian Ocean.[60]

Recent Developments in Shipbuilding

In 2009, the GoI formed an independent Shipping Ministry by bifurcating the erstwhile Ministry of Shipping, Road Transport and Highways. One of the major initiatives of the ministry is the formation of the Centre of Excellence in Maritime and Shipbuilding (CEMS). CEMS has 24 Laboratories.[61] These laboratories undertake skill development in ship hull design, ship detailed design, maintenance and repairs, and Life Cycle Management. The aim of the skill development initiative is to enable a seamless transition towards Industry 4.0.

Under the National Waterways Act of 2016, 106 new National Waterways (NWs) have been declared, for which it is estimated that the Inland Waterways Authority of India (IWAI) would require about 65 new vessels, and the Union Territory of Lakshadweep (UTL) would require about 14 new vessels by 2022-23.[62] However, Indian shipbuilding continues to be driven by defence requirements. There are 49 projects of the Indian Navy being executed by various shipyards, and about 50 projects are in the planning stage. There are

70 projects of the Indian Coast Guard which are presently under execution by various shipyards, and about 16 projects are at the planning stage.[63] The indigenisation plan of the Indian Navy is also having a positive impact on the rise of the ancillary equipment market. The requirement of vessels by the coastal waterways transportation is driving commercial shipbuilding. In 2018, IWAI placed an order for 10 ROPAX/RORO on CSL. Based on the Sagarmala Programme,[64] the southern state of Tamil Nadu has been identified as being developed as a maritime cluster. Gujarat Maritime Board (GMB) is developing a marine shipbuilding park in Bhavnagar, alongside a maritime services cluster in Ahmedabad – or Gujarat International Finance Tec-City (GIFT City).[65]

India is actively pursuing a policy of maritime cooperation in the Indian Ocean Region (IOR) which aims to provide Security And Growth for All in the Region (SAGAR).[66] The cooperation extends to its maritime neighbours, including Sri Lanka, Maldives, Mauritius, and Seychelles, in areas of trade, tourism, infrastructure, environment, Blue Economy, and security. India has enhanced its engagement with these countries by extending help in capacity building by providing radars, coastal surveillance equipment, vessels and aircraft, and establishing maritime infrastructure. The cooperation extends to White Shipping, Blue Economy, disaster response, anti-piracy and counter-terrorism, as well as hydrography.[67]

Another source of demand for the Indian shipbuilding industry is the construction of river-sea vessels, inland vessels, and fishing vessels. The industry is focussing on using emerging technologies, especially on constructing vessels that use alternate fuels. The GoI is encouraging public sector companies to source vessels required by them from Indian shipbuilders, in a bid to also attract the global suppliers of auxiliaries to set up plants in the country. India also has a fledging ship repair market; although presently, its market share is at 1 per cent of the global ship repair market, there is a huge growth potential since 7 to 9 per cent of global trade passes within 300 nautical miles of the coastline. A new dry dock is being constructed at CSL[68] to tap the market potential of building specialised and technologically advanced vessels, such as LNG carriers, Aircraft carriers of higher capacity, jack-up rigs, drill ships, larger dredgers, and repairing of offshore platforms and large vessels.[69] CSL is constructing 23 hybrid electric boats for the Kochi Water Metro and has signed a contract for the construction and supply of two autonomous electric ferries for ASKO Maritime AS, Norway, with an option to build two more identical vessels.[70]

NOTES

1   Radha Kumud Mookerji, *Indian Shipping: A History of the Sea-Borne Trade and Maritime Activity of the Indians from the Earliest Times*, (London: Longmans Green and Co., 1912), 1–2.

2   Sanyal, *Land of Seven Rivers*, ch.2, 70.

3   Sanyal, *Land of Seven Rivers*, 4–5.

4   The Dakshina Path or Southern Road ran from the Gangetic plains through Central India to the Southern tip of the peninsula.

5   The Uttara Path or Northern Road ran from eastern Afghanistan, through Punjab and Gangetic plains to the seaports of Bengal.

6   Sanyal, *Land of Seven Rivers*, ch.2, 70-86.

7   Romila Thapar, *The Penguin History of Early India: From the Origins to AD 1300*, (New Delhi: Penguin India, 2003), 236–37.

8   Sanyal, *Land of Seven Rivers*, ch.2, 116-18.

9   A Bronze Age civilisation lasting from 3300 BCE to 1300 BCE.

10   SR Rao, "Shipping and Maritime Trade of Indus People," *Expedition* vol.7, no. 3, (1965), 30-37.

11   Rajeshwar Nath, "Shipbuilding in India," *Indian Defence Review* vol. 22, no. 2, http://www.indiandefencereview.com/news/shipbuilding-in-india/, accessed 09 August 2020.

12   The Iron Age which refers to a period from about 1500 to 500 BCE.

13   Mookerji, *Indian Shipping*, ch.1, 37-38.

14   An ancient Indian Sanskrit treatise on statecraft, economic policy and military strategy authored by Kautilya who is also identified as Vishnugupta and Chanakya.

15   KM Panikkar, *India and the Indian Ocean* (London: Allen & Unwin, 1951), 31.

16   Mookerji, *Indian Shipping*, ch.1, 72-73.

17   Thapar, *The Penguin History of Early India*, ch.8, 236.

18   Mookerji, *Indian Shipping*, ch.1, 103.

19   V Kanakasabhai, *The Tamils Eighteen Hundred Years Ago* (New Delhi: Asian Educational Services, 1904), 34–35.

20   Mookerji, *Indian Shipping*, ch.1, 82-83.

21   Mamata Chaudhuri, "Ship-Building in the Yuktikalpataru and Samarangana Sutradhara," *Indian Journal of History of Science*, vol. 11, no. 2 (1976), 137–47.

22   Mookerji, *Indian Shipping*, ch.1, 14-19.

23   Nath, *Shipbuilding in India*, ch.8.

24   Krishnan, *Prosperous Nation Building Through Shipbuilding*, ch. 1, n. 5, 26–29.

25   Nath, *Shipbuilding in India*, ch.8.

26   Indians prepared fast dyes for textile fabrics through the treatment of natural dyes like 'manjishtha', with alum and other chemicals, and could extract indigotin from indigo plants.

27   Mentioned in Brihat-Samhita by Varahimihira in the 6th century AD.

28   Mookerji, *Indian Shipping*, ch.1, 128-29.

29   BK Apte, *A History of the Maratha Navy and Merchantships,* (Bombay: State Board for Literature and Culture, Govt. of Maharashtra, 1973), 110–11.

30   Mookerji, *Indian Shipping,* ch.1, 178-79.

31   RGS Cooper, *The Anglo-Maratha Campaigns and the Contest for India: The Struggle for Control of the South Asian Military Economy* (Cambridge: Cambridge University Press, 2007), 31.

32   K Sridharan, *Sea Our Saviour* (New Delhi: Taylor & Francis, 2000), 43.

33   Apte, *A History of the Maratha Navy and Merchantships,* ch.1, 110–11.

34   Ruttonjee Ardeshir Wadia, *The Bombay Dockyard and The Wadia Master Builders* (Bombay: Ruttonjee Ardeshir Wadia, 1955), 121.

35   Low Charles Rathbone, *History of the Indian Navy. (1613-1863) Volume 1,* vol. 1 (Online PDF copy, 1877), 297–99.

36   Ganesh Selvi, "Transportation Services and the World Trade Organization a Legal Study" (New Delhi, Jawaharlal Nehru University, 2009), http://hdl.handle.net/10603/14814, accessed 08 October 2020.

37   DG Tendulkar, *Mahatma Volume 3: Life of Mohandas Karamchand Gandhi*, (New Delhi: Publications Division Ministry of Information & Broadcasting, 1961), 99.

38   Nath, *Shipbuilding in India,* ch.8.

39   Radhe Shyam Rungta, *The Rise of Business Corporations in India 1851-1900* (Cambridge: Cambridge University Press, 1970), 291–92.

40   Daniel R Headrick, *The Tentacles of Progress: Technology Transfer in the Age of Imperialism, 1850-1940* (New York: Oxford University Press, 1988), 367.

41   Headrick, *The Tentacles of Progress,* 368–69.

42   Indian national shipping is considered to have been reborn on this day, and therefore, April 5 is celebrated every year as India's National Maritime Day.

43   Tata Iron and Steel Company (TISCO) was set up in Tatanagar in present-day Jharkhand in 1903.

44   HB Desai, "The Indian Shipping Industry with Special Reference to its Post War Developments" (Baroda, Maharaja Sayajirao University of Baroda, 1964), 219–24, http://hdl.handle.net/10603/58879, accessed 08 October 2020.

45   A regional organisation set up in 1951 for intergovernmental efforts to strengthen the economic and social development of member countries in the Asia-Pacific region.

46   "Cochin Shipyard Ltd.," https://cochinshipyard.com/history, accessed 04 October 2020.

47   "Planning Commission, Government of India: Five Year Plans," https://niti.gov.in/planningcommission.gov.in/docs/plans/planrel/fiveyr/index5.html, accessed 04 October 2020.

48   "Planning Commission, Government of India: Five Year Plans," https://niti.gov.in/planningcommission.gov.in/docs/plans/planrel/fiveyr/index5.html, accessed 04 October 2020.

49   NITI Aayog, "Strategy for New India @75," https://niti.gov.in/writereaddata/files/Strategy_for_New_India.pdf, accessed 22 October 2018.

50   GM Hiranandani, *Transition to Triumph: Indian Navy 1965-1975,* (New Delhi: Spantech & Lancer, 2009), 17.

51   FSA34 was a more advanced design that was under implementation by the Royal Navy ships under construction. It had a larger beam and catered for the installation of the latest weapon equipment.

52   Large numbers were associated with the British shipbuilders because in their philosophy construction was still an art and one had to learn on the job.

53   65 per cent of steel for the first frigate came from the Rourkela steel plant; later, the steel required — type B was produced by SAIL, and about 90 per cent of steel was sourced from indigenous sources.

54   Hiranandani, *Transition to Triumph,* ch.1, 102-5.

55   Southeast Asia Treaty organization was formed in 1954 with the US, UK, France, New Zealand, Australia, the Philippines, Thailand and Pakistan with the aim to prevent communism from gaining ground in Southeast Asia.

56   Central Treaty Organization was set up in 1955 to counter the threat of Soviet expansion in the Middle East oil-producing region. The member countries were Turkey, Iran, Pakistan, Iraq, the UK and later the US.

57   Hiranandani, *Transition to Triumph*, ch.1, 362-65.

58   GM Hiranandani, *Transition to Eminence: The Indian Navy 1976-1990* (New Delhi: Lancer International, 2009), 6.

59   Hiranandani, *Transition to Eminence*, 99.

60   GM Hiranandani, *Transition to Guardianship: The Indian Navy 1991–2000* (New Delhi: Lancer International, 2013), 125.

61   6 of these laboratories are in IRS Mumbai and 18 in IMU Visakhapatnam.

62   Ministry of Shipping, Govt. of India, *Annual Report 2017-18*, http://shipmin.gov.in/sites/default/files/2017-18%20English.pdf, accessed 06 October 2020. The achievement of the specified targets could, however, not be verified.

63   GK Harish, and Prashant Singh, "Shipbuilding: A Larger National Perspective," Vivekananda International Foundation, https://www.vifindia.org/sites/default/files/shipbuilding-a-larger-national-perspective.pdf, accessed 06 October 2020.

64   Ministry of Shipping, Government of India, *Projects under Sagarmala*, http://sagarmala.gov.in/projects/projects-under-sagarmala#:~:text=SagarMala%20%2D%20Projects%20 Under%20Sagarmala%20Ministry,Shipping%2C%20GOI%2C%20Government%20of%20India, accessed 06 October 2020.

65   Ministry of Shipping, Govt. of India, *Annual Report 2019-20*, http://shipmin.gov.in/sites/default/files/Shipping%20Annual%20Report%20English_compressed.pdf. accessed 06 October 2020.

66   Subhasish Sarangi, "Unpacking SAGAR (Security and Growth for All in the Region)," *United Service Institution of India*, https://usiofindia.org/wp-content/uploads/2020/02/USI-Occasional-Paper_2_19-Unpacking-SAGAR-Final-print-File-1.pdf, accessed 08 October 2020.

67   S Jaishankar, *The India Way: Strategies for an Uncertain World* (New Delhi: HarperCollins India, 2020), 166.

68   Graving Dock of Size 310 m × 75/60 m × 13 m and one gantry crane 600 Tons, Two 75 Tons LLTT cranes and other facilities.

69   Ministry of Shipping, Govt. of India, *Annual Report 2019-20*.

70   Ashish Singh, "Cochin Shipyard Signs Contract to Build Electric Vessels for Norwegian Company," *The Daily Guardian,* https://thedailyguardian.com/cochin-shipyard-signs-contract-to-build-electric-vessels-for-norwegian-company/, accessed 07 October 2020.

9

Conclusion

Evolution of Shipbuilding

Human settlements near water bodies, rivers, and seas found ways to utilise them for various needs, including livelihood, transportation, and trade. The medium of venturing into the waters evolved from boats, such as paddling and sailing canoes, galleys with oars, wooden sail ships, and engine-propelled steel ships to the present generation of autonomous unmanned vessels. Different types of vessels evolved around the world, customised to suit local requirements and ambitions.

A *tour de horizon* of shipbuilding reveals that in the extended Indian Ocean Region, there were three distinct traditions of shipbuilding – each within the geographical areas marked by ethnic and linguistic boundaries. The western half – as far as Bengal – followed an Indo-Islamic tradition, producing common hull shapes with vessels rigged with lateen sail. The Indonesian islands, Malaya, and Burma produced fast boats like Prahu and Sampan. The third group was the Chinese Junks which were massive rectangular barges. The Minoans, based in Crete, used sailing ships with oars, and the Egyptians built ships from papyrus and later used planks joined at ends. Shipbuilding in the Mediterranean evolved from technologies based on knowledge rooted in the works of ancient Greece and Rome.

Global trade relied on the logistic supply chains that were a mix of terrestrial routes and the seas. Around the tenth century AD, there was a substantial increase in inter-regional trade, and commodities exchanged at the inter-continental level began to be transported by sea rather than over land. Between the tenth and fifteenth centuries, Venice emerged as a trading hub in Europe for intra-Mediterranean maritime trade; inter-continental trade was carried out through the classical Silk Route. Chinese products were traded via the Caravan Routes to ports in the Black Sea, and Indian and other Asian products were transported via Syria and Alexandria.

Trade was managed in two distinct global patterns. Trade patterns in the Indian-Pacific Ocean Region were managed by commercial contracts that included established conventions, enforced by means of collective sanctions against any defaulting members. In the Mediterranean, however, from Graeco-Roman times and perhaps even earlier periods of history, dominant powers exercised control over the vital sea routes to control both economic resources and political settlements.

The fall of Constantinople in the mid-fifteenth century disrupted traditional trade routes, which were later restored through alternate routes established during the Great Voyages of the sixteenth century. The volume of sea commerce increased, and the two distinct regional trading systems of Europe and Asia were bridged, transplanting seedlings of the Mediterranean pattern into the Indian Ocean. This became possible as Europeans developed long-distance armed merchant shipping, floating fortresses, and warehouses, allowing them to expand their oceanic control from home bases in Europe and establish new bases in remote locations in Asia. Therefore, by the first decade of the sixteenth century, the period of peaceful sailing was over in the Indian Ocean, paving the way for an era where whoever controlled the sea was in the position of overwhelming commercial and political superiority; sea power became central to great power conflicts.

It is clear that as global trade integrated through the oceanic routes, two significant developments in mid-seventeenth-century Europe shaped the world order. Firstly, the concept of nation-states emerged as a result of the Treaty of Westphalia at the end of the Thirty Years' War. Secondly, the establishment of a three-tiered capitalist-based modern world economy – necessarily anarchic and dominated by Western Europe at the core. The Financial, Scientific, and Industrial Revolutions in the following centuries consolidated European dominance and centrality of sea power. Dominant world powers extensively employed colonialism, slavery, and enforced trading patterns. It now became a necessary requirement for a great power to be a major sea power, capable of dominating the seas for commercial and political superiority. By extension, sea power became a significant component of the comprehensive national power; shipbuilding capability and capacity were benchmarks for an aspiring great power, and sea powers were leading shipbuilding nations. Successive dominant powers, including the likes of the Venetians, Portuguese, Dutch, and British, were simultaneously great sea powers.

The emergence of new powers, such as Japan, Germany, and Italy in the latter half of the nineteenth century, challenging the established European powers and other contenders like the US and the USSR, eventually led to the two World Wars, resulting in a world order that was essentially bipolar. Shipbuilding continued to be a major factor during these transitions and great powers were necessarily major sea powers. The evolution of shipbuilding can best be summarised in the fishbone diagram in Figure 9.1.

Figure 9.1: Evolution of Shipbuilding

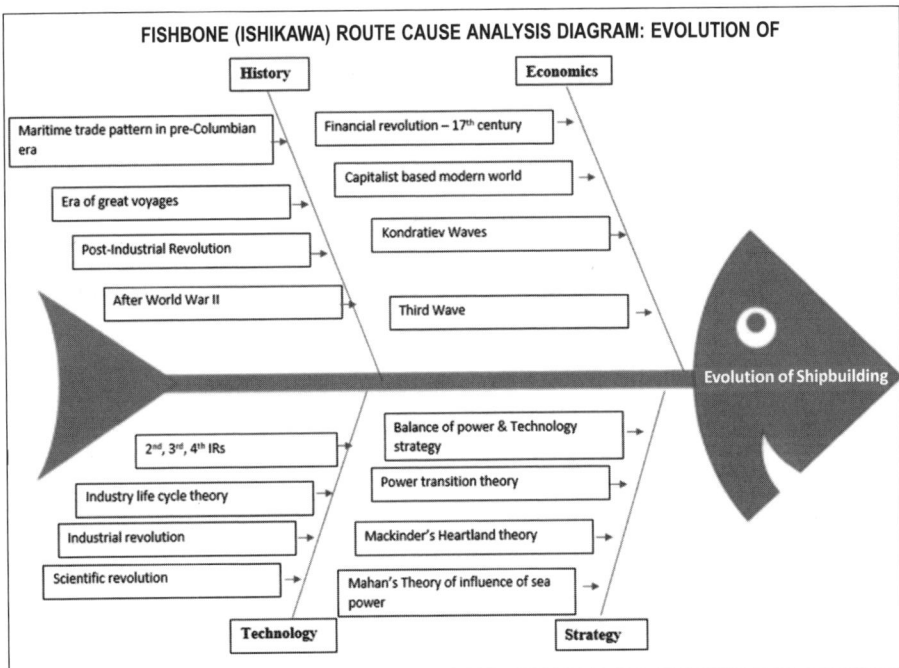

Shipbuilding Trends in the Indo-Pacific after World War II

It becomes clear that the multipolar world of 1885 was replaced by a bipolar world in 1945, at the end of World War II. The recalibration of great power equations had a significant impact on Asia, where – as the age of empires and colonies came to an end – many independent states were formed. From the American perspective, the growth and expansion of the Soviet Union seemed to validate the geopolitical predictions of Mackinder and others, which suggested that a massive military power would control the resources of the Eurasian heartland. Furthermore, the expansion of the State into the periphery or 'rimland' would have to be contested by the great maritime states to maintain a global balance of power.

In 1946, the US concluded that it enjoyed military superiority at sea and air, whilst the Soviet Union had significant supremacy on land. Considering their weakness on land within the Eurasian landmass, the Americans planned to restrict the use of force to those areas where the Soviet armies could be countered defensively by naval, amphibious, and air power. This strategic intent was to define the continued significance of sea power in the future great power rivalry.

The US policy of containment aimed to protect major centres of industrial power from Soviet expansion. These centres were identified as Western Europe and Japan. The core of American thinking was rooted in two lessons from its recent history. First, a belief that threats to political stability arise primarily from gaps between economic and social expectations and reality, hence the Marshall Plan. Second, the best protection against aggression is to possess overwhelming power and a political will to use it, which led to the establishment of the Atlantic Alliance.

The Marshall Plan was designed to get Europe on its feet economically and the North Atlantic Treaty Organization (NATO) was to look after its security. The Marshall Plan included massive economic assistance and the transfer of technologies to build infrastructure and industry in specific regions, such as Western Europe and Japan. Shipbuilding formed a major component of the transfer of technology to Japan. The Marshall Plan did not include the Korean Peninsula initially; however, the attack by North Korea on the South was seen as an expansion of communists beyond China, which needed to be contained. Consequently, the US entered the Korean War with decisive might and, subsequently, got involved in Vietnam.

In the context of technology power shifts, as countries in the Asia-Pacific region became the hotbed of reconstruction and growth, the shipbuilding industry became a major facilitator, since the US and the West were the leading shipbuilding nations. Since the shipbuilding industry is a manpower- and technology-intensive industry, it was most suited for transformation to an industrial ecosystem. The established shipbuilders of the US and Europe provided technology and governments made finances available and allowed market access to facilitate growth.

The US and its allies aided Japan in beginning the reconstruction of its industry with the intent of making it an anti-communism bastion in Asia. Since most of the pre-war shipbuilding infrastructure of Japan was largely intact, the shipbuilding industry grew rapidly with US assistance. The Suez Crisis in 1956 forced oil transportation from the Persian Gulf to the European nations to go around Africa, which required additional time and higher costs. To be able to transport larger quantities in each trip and achieve economies of scale, there was a huge demand for large tankers. At that time, European shipyards had the infrastructure to build ships smaller than 32,000 DWT, while only Japanese shipyards were constructing large tankers. These tankers were used to transport oil to Japan due to the absence of a canal in the transit route. The Japanese shipyards responded promptly to the market demand and made the best use of available low-cost and skilled manpower to capture the maximum global shipbuilding market share.

The OPEC Oil Crisis of 1973 – when crude production was reduced – substantially led to a drastic fall in the global demand for oil tankers. This triggered the Japanese shipbuilding industry's downturn, forcing reduced production and massive restructuring. The rise of shipbuilding in South Korea during this period, coupled with the availability of skilled manpower and favourable currency and steel costs, favoured a competitive shift toward South Korea. This was also facilitated by the US commitment to South Korea, following the end of the Korean War in 1953. By the 2000s, Korea emerged as a leader in shipbuilding.

The Chinese began to focus on shipbuilding during the Deng Xiaoping era in the mid-1970s. Initially, Chinese shipyards received technological aid and expertise from Japan and Korea. The availability of cheap labour in abundance; structural reforms in finance and management; and government policies to support production, exports, and manage demand collectively contributed to the growth of shipping as a strategic industry – leading to China attaining global leadership in 2009. The Chinese initially focused on lower-end technologies to increase their market share. Once they established themselves as a major player, they succeeded in forming consultancies with international majors in high-end shipbuilding, including LNG carriers and major international ancillary manufacturers.

The opening of markets and the focus on the shipbuilding sector in India gained momentum in the early nineties. Efforts have been made to channelise the infrastructure and technology to harness the abundant availability of cheap labour and boost the indigenous production of ancillary equipment. India's share in global shipbuilding, specifically offshore platform construction, has increased substantially in recent times, and its warship-building capability has grown substantially – with help from the Soviet Union as well as Europe.

It becomes clear that the emergence of Japan, Korea, China, and India, as well as the redistribution of the market share, have invariably been driven by the availability of abundant skilled manpower at a lower cost and the strategic management of technologies driving the shipbuilding industry. These maritime Asian countries had been great maritime powers throughout their history but had lost out in their shipbuilding capabilities – either due to their self-imposed isolation or colonial ingress. As these countries grew economically and industrialised, they established themselves as leading global shipbuilders. They also used their advanced merchant shipbuilding infrastructure to build ultra-modern warships and submarines and entered the global military-industrial complex. Therefore, the shipbuilding centre shifted from Europe and the US to maritime Asia, and the overall growth of these nations ushered in the Indo-Pacific construct.

The study reveals that, as the technologies of Industry 4.0 begin to dominate the markets, the countries in the Indo-Pacific are already established manufacturing hubs of the world and are competing in both products and innovations to dominate, furthering their significance. Shipbuilding, both in merchant as well as naval applications, is being transformed, and the countries in the Indo-Pacific are utilising their lead positions to consolidate and capture the market space in new products. The shipbuilding trends in the Indo-Pacific after the Second World War can best be summarised in the fishbone diagram in Figure 9.2.

Figure 9.2: Shipbuilding Trends in the Indo-Pacific

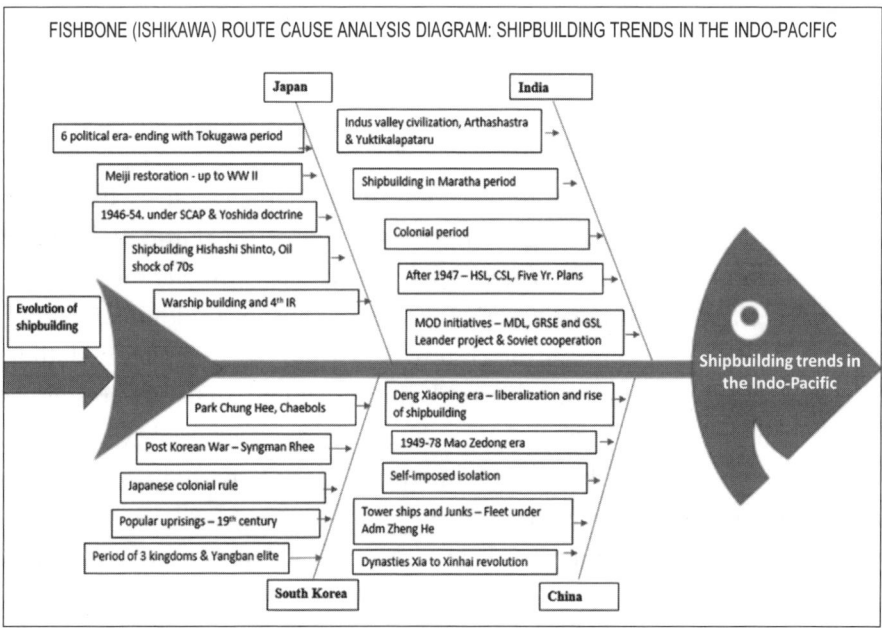

Correlation between the Shipbuilding Industry and Overall Growth

Studying the countries that experienced substantial economic growth at different periods of history establishes a positive correlation between the economic prosperity of a nation and the growth of shipbuilding. The correlation is stronger after the tenth century AD when global trade volumes increased through the seas. As the seas became a principal component of the global logistics supply chain and trade volumes increased, the demand for larger and more advanced ships increased over the centuries. Shipbuilding industries continued to grow due to higher demand, and leading nations became richer and more powerful. Shipbuilding involves many industrial trades and components, and therefore, major shipbuilding nations also became advanced industrial centres.

To secure sea trade, warships were deployed, and subsequently, the navies of leading nations were used – not only to protect their interests but also to expand their area of influence. The shipbuilding industry produced warships as well and experienced further growth. The regenerative growth relationship between shipbuilding, export-oriented industries, and sea power establishes a positive correlation between shipbuilding trends and the overall growth of a country.

The growing shipbuilding trend is also indicative of overall industrial growth. As the industrial output of a country grows, the country experiences an increase in exports, leading to higher volumes of sea traffic and greater access to global markets. The countries that experienced higher growth were those that effectively channelised their industrial output to dominate global markets. This is why most of the countries that were examined – including Portugal, the Netherlands, and the UK – exhibited that the growth of shipbuilding led to the overall growth of the countries as industry-based, export-oriented economies. Although shipbuilding was not the only contributor towards such growth, it was a prominent one. The trend has been consistent with maritime Asian countries, where each country that emerged as a major shipbuilding nation has also transformed into an export-oriented, industry-based economy.

Contribution of Shipbuilding in Maritime Asia

After World War II, the US and its allies began to rebuild some of the countries in Asia due to strategic compulsions. The process started with Japan, where the US provided finance and technologies, as well as access for Japanese goods to be sold in American markets. Amongst the areas of initial cooperation was the shipbuilding industry, since the pre-war shipbuilding infrastructure of Japan was largely intact. Japan rose to emerge as the world's largest shipbuilding nation in a decade. Since the shipbuilding industry also led to the setting up of many ancillary industries, there was overall industrial growth – including independently-driven sectors. Shipbuilding was a significant contributor to the overall economic growth of Japan. Similarly, other Asian countries that followed – including South Korea, China, and India, among others – gained from the technology transfers, investments, and market access made available by the West. The impact of this has been substantial, and as a result, the present collective market share of these countries is greater than ninety per cent of global total shipbuilding.

As shipbuilding contributed to the overall industrial and economic growth of the countries in maritime Asia, the collective share of the global Gross Domestic Product (GDP) from the region increased substantially. Economic growth also made it possible for these countries to raise powerful militaries and a supporting military-industrial complex. The magnitude and rapid economic growth of these countries have brought the Asia-Pacific Region to prominence, as it represents a convergence of major maritime trading and strategic powers that shape global events and contribute towards the formulation of the Indo-Pacific construct.

Maritime Civilisational Connect and the Indo-Pacific

Shipbuilding has been one of the most significant contributors to the overall growth of Asian countries after World War II. Many of these countries shared a maritime civilisational legacy and were great shipbuilding nations. However, shipbuilding in these countries did not grow and transform with the Industrial Revolution, resulting in the countries completely missing out on the transition from the wood and sail era to the steel and steam era. Japan undertook self-imposed isolation during the 250 years of Tokugawa Rule, as did South Korea during the sixteenth and seventeenth centuries, and China during the Ming Dynasty in the fifteenth century. Indian shipbuilding declined during colonial rule. Therefore, it becomes clear that the modem shipbuilding capability of maritime Asian countries – which contributed significantly to the Indo-Pacific construct – is largely based on the transfer of technologies from the West rather than a continuity of the Asian civilisational maritime technological culture.

In Summary

Looking at the broad pattern that emerges from the study, it becomes clear that after World War II, the US and its Allies began to rebuild Japan as part of the Marshall Plan. The strategic intent was to contain Soviet expansion, while the execution methodology involved making finances available to Japan, facilitating the transfer of modern technologies, and providing access to American markets. One of the major industries in which the transfer of technology began was shipbuilding. This was because the US was the world leader in shipbuilding, and at the end of the war, it had tremendous excessive capacity as well as falling demand. The shipbuilding industry was a matured industry as per the industry life cycle theory, and was most suitable for relocation to destinations where there was availability of low-cost skilled workforce. Japan had the advantage that most of its pre-war shipbuilding infrastructure was intact, thus, shipbuilding – being a technology- and manpower-intensive industry – was most suitable for transformation to an industrial society.

In 1946, Japan's Kure Shipyard was leased to the National Bulk Carriers Inc. of New York for constructing a large iron ore vessel – on condition that the company would use Japanese steel and provide access to Japanese engineers for the transfer of technology. In the ensuing decade, financial aid, transfer of technology, access to American markets, and above all, the Japanese culture of hard work and perfection inherited from the Tokugawa Period resulted in Kure Shipyard emerging as the most technically advanced facility of its kind in the world. It became a reference for shipyards across Japan, producing high-quality ships at competitive prices. Japanese shipyards grew exponentially, and Japan became the world's largest producer of merchant ships by 1956. In a structured manner, Japan implemented agricultural reforms, built up industrial infrastructure, and set up an export-oriented economy. In the first decade, Japan focused on steel production and heavy industries like shipbuilding; in the 1960s and 1970s, it focused on consumer products and automobiles for export, and later developed knowledge-based products like computers and electronics.

Between 1965 and 1995, South Korea, Taiwan, Hong Kong, and Singapore – often referred to as the Four Tigers – raised per capita incomes six-fold. During the same period, Indonesia, Malaysia, and Thailand tripled their income levels. The export-led industrialisation of South Korea was facilitated by substantial aid from the US after the Korean War, and subsequently from Japan in the 1960s. It focussed on achieving an export-driven economy, with heavy investment in technology-intensive industries – such as shipbuilding, steel, petrochemicals, and electronics. The US normalised its relations with China in 1972, opening channels for American investments. China implemented reforms and, from 1978, opened its markets under Deng Xiaoping. The commercial shipbuilding output of China surged thirteen-fold between 2002 and 2012, and it became the world's largest shipbuilder by 2015. India implemented important reforms in the 1990s by opening its economy. India's growth has been significantly diversified with the service sector recording the highest growth, followed by the industry.

Whilst the West, led by the US, facilitated the rise of export-oriented industrial growth of countries in maritime Asia, they took great care to apply technology denial regimes in military and dual-role technologies. Therefore, the West continued its dominance of the military-industrial complex. In the context of shipbuilding, the West continued to have control and took the lead in weapon systems and components of military specifications ('mil specs'), as well as warship and submarine design and production. Consequently, in the period after World War II, while many maritime Asian countries emerged as leading shipbuilding nations of the world, they were not great sea powers. Until contemporary history, a great sea power was also a leading shipbuilding nation. Some of the developments that followed changed this situation.

The maritime Asian countries which became global leaders in merchant shipbuilding, began to utilise their advanced shipbuilding infrastructure towards building modern warships and submarines. Although this started with the intent of indigenising warship production and supporting their own shipbuilding companies with orders when there was a slump in demand for merchant ships, it provided them with an opportunity to enter the military-industrial complex and export warships. Some of the leading European and American warship-producing shipyards collaborated with the shipyards in these countries since they were short of orders and needed business to sustain. However, the weapons and electronics continued to be sourced from leading Western companies. This trend changed due to major developments in the consumer electronics market as well as changes in the world order.

In the 1980s, there was an exponential growth in Information Technology and telecommunications, fuelled by demand in the consumer market. The electronics market, including microelectronics and the printed circuit board industry, had been dominated by the defence industry's requirements until now. The rise in product demand and sales fuelled more investments by the commercial industry into research and development. The electronics, networking, and software sectors experienced unprecedented growth, leading to a substantial increase in market size. This growth fuelled competition and companies ruggedised their products to cut costs, ensure low failure rates, and reduce warranty returns in order to broaden their loyal customer base. Product ruggedisation also ensured lesser wastage during transportation. So, there emerged a pool of products which were rugged, competitively-priced (as compared to mil-spec variety), and current in technology. This made it attractive for defence industries to use commercial off-the-shelf (COTS) products as building blocks rather than pursue exclusivity in design. Many of these products were mass-produced in the maritime Asian manufacturing hub.

Transitions in the world order – including the disintegration of the Soviet Union and the unification of Germany – led to a reduction in defence budgets around the world, forcing defence planners to cut down expenditures. As a direct consequence, the defence industry came under tremendous pressure to cut down the costs of systems under development. There was also an urgent requirement for defence forces to upgrade legacy systems to extend their life. This was imperative to maintain the force levels, and the cost involved for this alone was tremendous for large- and medium-size navies. COTS-based solutions appeared to be the best option with the initial proliferation of electronics, information technologies, and networking sectors.

The countries in maritime Asia – which, by now, were the manufacturing hub of the world and leading producers in many sectors, including merchant shipbuilding – entered the military-industrial complex, building state-of-the-art warships and submarines. Whilst the West continues to make the most advanced ships and submarines, the cost of production is much higher when compared to a similar capability warship built in one of the shipyards in Asia. The proliferation of COTS in military systems, along with collaboration between Western shipyards and leading shipbuilders in maritime Asia, has limited the impact of technology denial regimes, which were pursued more effectively in the past. The collective effect of these changes has led to the emergence of the Indo-Pacific construct as a significant factor in shaping global developments.

The technologies of Industry 4.0 are bringing about significant changes across all sectors, including merchant and warship building. The significance of sea power and the global logistics supply chains – through the seas – retain their importance while undergoing huge transformations with the applications of Industry 4.0. Shipyard 4.0 is extensively using the Internet of Things, Artificial Intelligence, Augmented/Virtual Reality, and 3D printing in construction and designs. Moreover, the shipping sector is using digital twinning, blockchain technology, AI, and AR/VR to improve efficiency and ensure on-time deliveries at the lowest price, while the navies are looking at unmanned warships and submarines, 3D-printed spares and drones onboard, blockchain technology in tactical networks, etc. Industry 4.0 has been on the rise since the early 2000s when the manufacturing hub was already well-established in Indo-Pacific countries. This has enabled the countries of the region to emerge as serious contenders for the world's lead position; hence, shipbuilding continues to be an integral element of the Indo-Pacific construct.

BIBLIOGRAPHY

Reports/Papers/Documents

"Cochin Shipyard Ltd.," https://cochinshipyard.com/history.

Department of Defence, United States of America, *The Asia Pacific Maritime Security Strategy,* https://dod.defense.gov/Portals/1/Documents/pubs/NDAA%20A-P_Maritime_SecuritY_Strategy-08142015-1300.pdf.

"Indian Naval Indigenisation Plan (INIP) 2015-2030", http://indiannavy.nic.in/sites/default/themes/indiannavy/images/pdf/naval_initiatives/INIP_2015-2030.pdf.

Ministry of Shipping, Government of India, *Annual Report,* 2014-15, http://shipmin.gov.in/publication/annual-reports.

Ministry of Shipping, Government of India, *Maritime Agenda 2010-2020,* https://pib.gov.in/newsite/PrintRelease.aspx?relid=69044.

Ministry of Shipping, Government of India, *Annual Report,* 2017-18, http://shipmin.gov.in/publication/annual-reports.

Ministry of Shipping, Government of India, *Annual Report,* 2019-20, http://shipmin.gov.in/publication/annual-reports.

Ministry of Shipping, Government of India, *Projects under Sagarmala,* http://sagarmala.gov.in/projects/projects-under-sagarmala#:~:text=Sagar Mala%20%2D %20Projects%20Under%20Sagarmala%20Ministry,Shipping%2C%20GO I%2C%20Government%20of%20India.

NITI Aayog, *Strategy for New India @75,* https://niti.gov.in/writereaddata/files/Strategy_for_New_India.pdf.

Office of the Historian, Bureau of Public Affairs, United States Department of State, *The Washington Naval Conference, 1921–1922,* https://history.state.gov/milestones/1921-1936/naval-conference.

"Planning Commission, Government of India: Five Year Plans," https://niti.gov.in/planningcommission.gov.in/docs/plans/planrel/fiveyr/index5.html.

United Nations Conference on Trade and Development, UNCTAD/RMT/2014, United Nations Publication, *Review of Maritime Transport 2014,* https://unctad.org/system/files/official-document/rmt2014_en.pdf.

World Bank, India Development Update, March 2018, *India's Growth Story,* https://doi.org/10.1596/29515.

Books

Amsden, Alice H. *Asia's Next Giant: South Korea and Late Industrialization.* New York: Oxford University Press, 1989.

Anthony, Slaven. *The Shipbuilding Industry: A Guide to Historical Records.* Edited by L. A. Ritchie. Manchester, England: Manchester University Press, 1992.

Apte, Dr BK. *A History of the Maratha Navy and Merchantships.* Bombay: State Board for Literature and Culture, Govt. of Maharashtra, 1973.

Blume, Kenneth J. *Historical Dictionary of the U.S. Maritime Industry.* 1st edition. Plymouth, UK: Scarecrow Press, 2011.

Bowers, Ian. *The Modernisation of the Republic of Korea Navy: Seapower, Strategy and Politics.* New York: Springer, 2018.

Braudel, Fernand. *Civilization and Capitalism, 15th-18th Century, Vol. I: The Structure of Everyday Life.* Translated by Siân Reynold. Berkeley: University of California Press, 1992.

Bull, Hedley, and Adam Watson. *The Expansion of International Society.* Oxford: Oxford University Press, 1985.

Butterfield, Herbert. *The Origins of Modern Science 1300-1800.* Revised edition. London: Free Press, 1959.f

Chaudhuri, KN. *Trade and Civilisation in the Indian Ocean: An Economic History from the Rise of Islam to 1750.* Cambridge: Cambridge University Press, 1985.

Cooper, Randolf G. S. *The Anglo-Maratha Campaigns and the Contest for India: The Struggle for Control of the South Asian Military Economy.* Cambridge: Cambridge University Press, 2007.

Crowther, J. G. Whiddington, R. *Science at War.* London: His Majesty's Stationary Office, 1947.

Cumings, Bruce. *Korea's Place in the Sun: A Modern History* (Updated Edition). New York: W. W. Norton & Company, 2005.

Davies, Peter. Japanese *Shipping and Shipbuilding in the Twentieth Century: The Writings of Peter N. Davies.* Folkestone, UK: BRILL, 2009. http://ebookcentral.proquest.com/lib/inain/detail.action?docID=849000.

Diamandis, Peter H and Steven Kotler. *Bold: How to Go Big, Create Wealth and Impact the World*. Reprint edition. New York: Simon & Schuster, 2015.

Dickson, PGM. *The Financial Revolution in England: A Study in the Development of Public Credit, 1688-1756*. New York: Routledge, 2017.

Dreyer, Edward L. *Zheng He: China and the Oceans in the Early Ming Dynasty, 1405-1433*. 1 edition. New York: Pearson, 2006.

Dunkley, Mark. *Ships and Boats: 1840 - 1950*. Edited by Paul Stamper. Historic England, 2016. https://content.historicengland.org.uk/images-books/publications/iha-ships-boats-1840-1950/heag133-ships-and-boats-1840-1950-iha.pdf/.

Erickson, Andrew S., ed. *Chinese Naval Shipbuilding: An Ambitious and Uncertain Course*. Annapolis, Maryland: Naval Institute Press, 2017.

Falk, Kevin L. *Why Nations Put to Sea: Technology and the Changing Character of Sea Power in the Twenty-First Century*. New York: Routledge, 2018.

Geanakoplos, Deno John. *Constantinople and the West: Essays on the Late Byzantine (Palaeologan) and Italian Renaissances and the Byzantine and Roman Churches*. Wisconsin: Univ of Wisconsin Press, 1989.

Griffis, William Elliot. *Corea the Hermit Nation: I.—Ancient and Mediaeval History. II.—Political and Social Corea. III—Modern and Recent History*. New York: C. Scribner's sons, 1888.

Harari, Yuval Noah. *Sapiens*. London: Penguin Random House, 2015.

Headrick, Daniel R. *The Tentacles of Progress: Technology Transfer in the Age of Imperialism, 1850-1940*. New York: Oxford University Press, 1988.

Hiranandani, GM. *Transition to Eminence: The Indian Navy 1976-1990*. New Delhi: Lancer International, 2009.

Hiranandani, GM. *Transition to Triumph: Indian Navy 1965-1975*. New Delhi: Spantech & Lancer, 2009.

Hiranandani, GM. *Transition to Guardianship: The Indian Navy 1991–2000*. New Delhi: Lancer Publishers LLC, 2013.

Hughes, Christopher W. *Japan's Re-Emergence as a "Normal" Military Power*. New York: Routledge, 2006.

Jacobson, Annie. *Operation Paperclip- The Secret Intelligence Program That Brought Nazi Scientists to America*. New York: Little, Brown and Company, 2014.

Jaishankar, S. *The India Way: Strategies for an Uncertain World*. New Delhi: Harper Collins India, 2020.

K Sridharan. *Sea Our Saviour*. New Delhi: Taylor & Francis, 2000.

Kanakasabhai, V. *The Tamils Eighteen Hundred Years Ago*. New Delhi: Asian Educational Services, 1904.

Kennedy, Paul. *The Rise and Fall of the Great Powers*. Great Britain: Fontana Press, 1989.

Kim, Joungwon Alexander. *Divided Korea: The Politics of Development, 1945-1972*. Elizabeth, NJ: Hollym Intl, 1999.

Kirchberger, Sarah. *Assessing China's Naval Power: Technological Innovation, Economic Constraints, and Strategic Implications*. New York: Springer, 2016.

Kissinger, Henry. *Diplomacy*. Reprint edition. New York, NY: Simon & Schuster, 1995.

———. *On China*. India: Penguin Books India Pvt Ltd, 2012.

———. *World Order*. Reprint edition. New York: Penguin Books, 2015.

Krishnan, S Navaneetha. *Prosperous Nation Building Through Shipbuilding*. New Delhi: KW Publishers Pvt Ltd, 2013.

Landes, David S. *The Unbound Prometheus: Technological Change and Industrial Development in Western Europe from 1750 to the Present*. Cambridge: Cambridge University Press, 2003.

Lane, Frederic Chapin, Arthur Donovan, Blanche D. Coll, Gerald J. Fischer, and David B. Tyler. *Ships for Victory: A History of Shipbuilding under the U.S. Maritime Commission in World War II*. Johns Hopkins Paperbacks Ed edition. Baltimore: Johns Hopkins University Press, 2001.

Lee, Jongsoo James. *The Partition of Korea After World War II: A Global History*. 2006 edition. New York, N.Y.: Palgrave Macmillan, 2006.

Lie, John. *Han Unbound: The Political Economy of South Korea*. Stanford Calif: Stanford University Press, 1998.

Lorge, Peter. *War, Politics and Society in Early Modern China, 900-1795*. New York: Taylor & Francis e-Library, 2006.

Lyon, David. *The Ship: Steam, Steel and Torpedoes: The Warship in the 19th Century*. First Edition. London: Her Majesty's Stationary Office, 1980.

Mackinder, HJ. *The Geographical Pivot of History*. Oxford: Blackwell Publishing, Ltd. 1904.

Maddison, Angus, Organization for Economic Co-operation and Development, and Organization for Economic Cooperation and Development. *The World Economy: A Millennial Perspective*. Paris, France: Organization for Economic Co-operation and Development, 2001.

Mahan, Alfred Thayer. *Influence of Sea Power Upon the French Revolution and Empire, 1793-1812* vol. 2, 1897. http://community.ebooklibrary.org/eBooks/

WPLBN0000873225-Influence-of-Sea-Power-Upon-the-French-Revolution-and-Empire-1793-1812—Vol-2-by-Mahan-A-T—Alfred-Thayer-.aspx?&Words=A%20T%20Mahan.

———. *The Influence of Sea Power Upon History, 1660-1783*, 1918. http://community.ebooklibrary.org/eBooks/WPLBN0000870131-The-Influence-of-Sea-Power-Upon-History-1660-1783-by-Mahan-A-T—Alfred-Thayer-.aspx?&Words=A%20T%20Mahan.

Massie, Robert K. *Dreadnought: Britain, Germany, and the Coming of the Great War.* Reprint edition. New York: Ballantine Books, 1992.

Meyer, Milton W. *Japan: A Concise History.* Lanham, MD, United States: Rowman & Littlefield Publishers, 2012. http://ebookcentral.proquest.com/lib/inflibnet-ebooks/detail.action?docID=467188.

Milewski, John O. *Additive Manufacturing of Metals: From Fundamental Technology to Rocket Nozzles, Medical Implants, and Custom Jewelry.* Switzerland: Springer, 2017.

Mookerji Radha Kumud. *Indian Shipping: A History of the Sea-Borne Trade and Maritime Activity of the Indians from the Earliest Times.* London: Longmans Green and Co., 1912. http://archive.org/details/in.ernet.dli.2015.237225.

Moore, Thomas G. *China in the World Market: Chinese Industry and International Sources of Reform in the Post-Mao Era.* Cambridge: Cambridge University Press, 2002. http://ebookcentral.proquest.com/lib/inain/detail.action?docID=202025.

Munoz, Joseph Mark S., ed. *Handbook on the Geopolitics of Business.* Cheltenham, UK: Edward Elgar Publishing Ltd, 2013.

Nichols, Wallace J., and Celine Cousteau. Blue Mind: *The Surprising Science That Shows How Being Near, In, On, or Under Water Can Make You Happier, Healthier, More Connected, and Better at What You Do.* New York: Little, Brown and Company, 2014.

Nye, Jr., Joseph S. *The Future of Power.* United States: Public Affairs, 2011.

Palais, James B. *Politics and Policy in Traditional Korea.* Cambridge: Harvard University Press, 1975.

Panikkar, KM. *India and the Indian Ocean.* London: Allen & Unwin, 1951.

Park, Chung Hee. *The Country, the Revolution and I.* Seoul: Hangchun-Dan Mansion, 1963.

Possony, Stefan Thomas, J. E. Pournelle, and Francis X. Kane. *The Strategy of Technology.* Electronic Edition. WebWrights, 1997. https://www.jerrypournelle.com/slowchange/Strat.html.

Preston, Antony. *The Ship. Dreadnought to Nuclear Submarine.* London: Her Majesty's Stationary Office, 1980. https://www.abebooks.co.uk/first-edition/ Ship-Dreadnought-Nuclear-Submarine-Antony-Preston/645553868/bd.

Rathbone, Low Charles. *History of the Indian Navy. (1613-1863)* Volume 1. Vol. 1. 2 vols. Online PDF copy, 1877.

Rungta, Radhe Shyam. *The Rise of Business Corporations in India 1851-1900.* Cambridge: Cambridge University Press, 1970.

Ruttonjee Ardeshir Wadia. *The Bombay Dockyard and The Wadia Master Builders. Bombay:* Ruttonjee Ardeshir Wadia, 1955.

Sakhuja, Vijay. *Asian Maritime Power in the 21st Century: Strategic Transactions: China, India and Southeast Asia.* Singapore: ISEAS, 2011.

Saliba, George. *Islamic Science and the Making of the European Renaissance.* Massachusetts: MIT Press, 2007.

Sanyal, Sanjeev. *Land of Seven Rivers - A Brief History of India's Geography.* Gurgaon: Penguin Books India Pvt Ltd, 2013.

Schumpeter, Joseph A. *Capitalism, Socialism and Democracy.* London: Routledge, 2013.

Schwab, Klaus. *The Fourth Industrial Revolution.* New York: Portfolio Penguin, 2017.

Seth, Michael J. *A History of Korea: From Antiquity to the Present.* Lanham, Md: Rowman & Littlefield Publishers, 2010.

Shi, Zhongzhi. *Advanced Artificial Intelligence.* Singapore: World Scientific Publishing Co Pte Ltd, 2014. http://ebookcentral.proquest.com/lib/inflibnet-ebooks/detail.action?docID=840558.

Soares, Carlos Guedes, Roko Dejhalla, and Dusko Pavletic. *Towards Green Marine Technology and Transport.* New York: CRC Press, 2015.

Studwell, Joe. *How Asia Works: Success and Failure in the World's Most Dynamic Region.* New York: Grove Press, 2013.

Tendulkar, D. G. *Mahatma Volume 3: Life of Mohandas Karamchand Gandhi.* New Delhi: Publications Division Ministry of Information & Broadcasting, n.d.

Thapar, Romila. *The Penguin History of Early India: From the Origins to AD 1300.* First edition. New Delhi: Penguin India, 2003.

Thiesen, William H. *Industrializing American Shipbuilding: The Transformation of Ship Design and Construction, 1820-1920.* Gainesville: University Press of Florida, 2006.

Todd, Daniel. *Industrial Dislocation: Case of Global Shipbuilding*. New York: Routledge, 1991.

Todd, Daniel, and Michael Lindberg. *Navies and Shipbuilding Industries: The Strained Symbiosis*. Westport: Praeger Publishers, 1996.

Toffler, Alvin. *The Third Wave*. New York: Bantam Books, 1981.

Turner, Colin, and Debra Johnson. *Global Infrastructure Networks: The Trans-National Strategy and Policy Interface*. Cheltenham: Edward Elgar Publishing, 2017.

Wallerstein, Immanuel Maurice. *The Essential Wallerstein*. New York: New Press, 2000.

Wijnolst, N and Tor Wergeland. *Shipping Innovation*. Washington: IOS Press, 2009.

Yeung, Henry Wai-chung. *Strategic Coupling: East Asian Industrial Transformation in the New Global Economy*. New York: Cornell University Press, 2016.

Journals/Magazines/Newspaper Articles

"2015-Marine-Industries-Resource-Guide-Japan-and-China.Pdf." Accessed 15 July 2019. http://www.mitc.com/wp-content/uploads/2015/04/2015-Marine-Industries-Resource-Guide-Japan-and-China.pdf?3dc2e8.

Abramowski, Tomasz. "Application of Artificial Intelligence Methods to Preliminary Design of Ships and Ship Performance Optimization." *Naval Engineers Journal* vol. 125, n. 3 (September 2013): 101–12.

Agarwala, Nitin, and Rana Divyank Chaudhary. "Growth of Shipbuilding in China: The Science, Technology, and Innovation Route." *Institute of Chinese Studies, Delhi*, (Occasional Paper, 31 May 2019): 29.

Alger, Leah. "Hyundai HI to Use Robots for Shipbuilding." *Software Testing News*, 15 May 2018. https://www.softwaretestingnews.co.uk/hyundai-hi-to-use-robots-for-shipbuilding/.

Altbach, Philip G. "The Past, Present, and Future of the Research University." *Economic and Political Weekly* vol. 46, no. 16 (April 2011): 65–73.

Ayres, Robert U. "Technological Transformations and Long Waves. Part I." *Technological Forecasting and Social Change* vol. 37, no. 1 (01 March 1990): 1–37. https://doi.org/10.1016/0040-1625(90)90057-3.

———. "Technological Transformations and Long Waves. Part II." *Technological Forecasting and Social Change* vol. 37, no. 2 (01 April 1990): 111–37. https://doi.org/10.1016/0040-1625(90)90065-4.

Bagchi, Indrani. "Peaceful Periphery On The Seas - Why 'Indo-Pacific' Is Nudging

'Asia-Pacific' off the Table Even as China's Shadow Looms." *Times of India*. 29 May 2018.

Baker, Glenn E, Richard A Boser, and Daniel L Householder. "Coping at the Crossroads: Societal and Educational Transformation in the United States." *Journal of Technology Education* vol. 4, no. 1 (1992).

Barbour, Violet. "Dutch and English Merchant Shipping in the Seventeenth Century." *The Economic History Review* vol. 2, no. 2 (January 1930): 261–90.

Baroch P. "International Industrialization Levels from 1750 to 1980." *Journal of European Economic History* vol. 11, n. 2, (Fall 1982): 269.

Bellias, Matt. "The Evolution of Maintenance towards Prescriptive." *Maintenance evolution prescriptive*, 14 March 2017. https://www.ibm.com/blogs/internet-of-things/maintenance-evolution-prescriptive/.

Bitzinger, Richard A. "S Korean Naval Shipbuilding: Full Speed Ahead." *Asia Times*, 30 June 2019. https://asiatimes.com/2019/07/s-korean-naval-shipbuilding-full-speed-ahead/.

Bloomberg. "At This Farm, the Boss Is an AI-Powered Algorithm." *The Economic Times*. 21 September 2018. https://economictimes.indiatimes.com/small-biz/startups/newsbuzz/at-this-farm-the-boss-is-an-ai-powered-algorithm/articleshow/65895095.cms.

Carter, Sarah. "AI-Powered Maritime Applications Launched by Transas." *ShipInsight*, 31 January 2018. https://shipinsight.com/articles/ai-powered-maritime-applications-launched-transas.

Caudell, T P, and D W Mizell. "Augmented Reality: An Application of Heads-up Display Technology to Manual Manufacturing Processes." *Proceedings of the Twenty-Fifth Hawaii International Conference on System Sciences*, vol. 2 no. 2 (1992): 659–69. https://doi.org/10.1109/HICSS.1992.183317.

Chauhan, Abhishek, Yogesh Kumar, Siddartha Mashetty, Anirban Bhattacharyya, and Om Prakash Sha. "A Machine Learning-Based Approach to Predict Corrosion Allowance for Ships." *International Society of Offshore and Polar Engineers,* 30 July 2018. https://www.onepetro.org/conference-paper/ISOPE-I-18-442.

Chen, Stephen. "China Developing Robotic Subs to Launch New Era of Sea Power." *South China Morning Post*, 22 July 2018. https://www.scmp.com/news/china/society/article/2156361/china-developing-unmanned-ai-submarines-launch-new-era-sea-power.

Chida, Tomohei, and Peter N Davies. "The Japanese Shipping and Shipbuilding

Industries." *International Journal of Maritime History* vol. 2, no. 2 (01 December 1990): 241–45. https://doi.org/10.1177/084387149000200213.

Chong, Key Ray. "The Tonghak Rebellion: Harbinger of Korean Nationalism." *Journal of Korean Studies (1969-1971)* vol. 1, no. 1 (1969): 73–88.

Chaudhuri, Mamata. "Ship-Building in the Yuktikalpataru and Samarangana Sutradhara." *Indian Journal of History of Science* vol. 11, no. 2 (1976): 137–47.

Chowdhury, Hasan. "Liver Success Holds Promise of 3D Organ Printing." *Financial Times,* 05 March 2018. https://www.ft.com/content/67e3ab88-f56f-11e7-a4c9-bbdefa4f210b.

Colton, Tim, and La Var Huntzinger. "A Brief History of World Shipbuilding in Recent Times." *The CNA Corporation* (September 2002). http://www.dtic.mil/dtic/tr/fulltext/u2/a409101.pdf.

Corner, Stuart. "NZ Navy First with IoT-Equipped Warship." *Computerworld.* 10 March 2017. https://www.pcworld.idg.com.au/article/615707/nz-navy-first-iot-equipped-warship/.

Crowley, Roger. "Arsenal of Venice: World's First Weapons Factory." *Military History* vol. 27, n. 6, (March 2011): 62-70.

"CSSC-CSIC Megamerger Confirmed at Last." *The Maritime Executive.* 01 July 2019. https://maritime-executive.com/article/cssc-csic-shipbuilding-megamerger-confirmed-at-last.

Daniels, Guy. "Inmarsat Introduces Smart Shipbuilding as Telcos Look to IoT Service Opportunities." *TelecomTV,* 05 September 2017. https://www.telecomtv.com/content/iot/inmarsat-introduces-smart-shipbuilding-as-telcos-look-to-iot-service-opportunities-15924/.

Das, Sumit, Aritra Dey, Akash Pal, and Nabamita Roy. "Applications of Artificial Intelligence in Machine Learning: Review and Prospect." *International Journal of Computer Applications* vol. 115, no. 9 (2015): 31-41.

D'Aveni, Richard A. "3-D Printing Will Change the World." *Harvard Business Review,* 01 March 2013. https://hbr.org/2013/03/3-d-printing-will-change-the-world.

Desai,H B. "The Indian Shipping Industry with Special Reference to Its Post War Developments." *Maharaja Sayajirao University of Baroda,* (1964). http://hdl.handle.net/10603/58879.

Drennan, Jimmy. "Harnessing Tech Innovation from Blockchain to Kill Chain." *Center for International Maritime Security,* 11 July 2018. http://cimsec.org/harnessing-tech-innovation-from-blockchain-to-kill-chain/36996.

Erickson, Andrew S. "Chinese Naval Shipbuilding: Full Steam Ahead." *The Maritime Executive,* 18 January 2019. https://www.maritime-executive.com/editorials/chinese-naval-shipbuilding-full-steam-ahead.

Erickson, Andrew S. Interview by Tiezzi Shannon. "Chinese Naval Shipbuilding: Measuring the Waves." *The Diplomat.* 25 April 2017. https://thediplomat.com/2017/04/chinese-naval-shipbuilding-measuring-the-waves/.

Evans, Gareth. "The Digital Naval Shipyard." *Naval Technology,* 12 February 2018. https://www.naval-technology.com/features/digital-naval-shipyard/.

Feldman, Leslie. "Cyber Security and the Fourth Industrial Revolution." *Symantec Corporation* (blog), 28 April 2018. https://www.symantec.com/blogs/expert-perspectives/cyber-security-and-fourth-industrial-revolution.

Florence, Hourtouat. "Peer Review of the Korean Shipbuilding Industry and Related Government Policies." *OECD Report C/WP6(2014)10.* Paris, France: OECD, January 2015.

Fraga-Lamas, P, Fernández-Caramés, T M, Ó Blanco-Novoa, and Vilar-Montesinos M A. "A Review on Industrial Augmented Reality Systems for the Industry 4.0 Shipyard." *IEEE Access* vol. 6 (21 February 2018). 13358–75. https://doi.org/10.1109/ACCESS.2018.2808326.

Frazer, Kelly A. "Decoding the Human Genome." *Genome Research* vol. 22, no. 9 (01 September 2012), 1599–1601. https://doi.org/10.1101/gr.146175.112.

Frith, Jake. "'Digital Twins' Approach Could Cut Costs in Shipbuilding." *Maritime Journal,* 06 November 2018. https://www.maritimejournal.com/news101/vessel-build-and-maintenance/ship-and-boatbuilding/digital-twins-approach-could-cut-costs-in-shipbuilding.

Gady, Franz-Stefan. "Japan Commissions 10th Soryu-Class Diesel-Electric Attack Submarine." *The Diplomat,* 27 March 2019. https://thediplomat.com/2019/03/japan-commissions-10th-soryu-class-diesel-electric-attack-submarine/.

Gambetta, Jay M., Chow, Jerry M, and Matthias, Steffen. "Building Logical Qubits in a Superconducting Quantum Computing System." *Npj Quantum Information* vol. 3, no. 1 (13 January 2017), 2. https://doi.org/10.1038/s41534-016-0004-0.

Ganesh, Selvi. "Transportation Services and the World Trade Organization a Legal Study." Jawaharlal Nehru University, 2009. http://hdl.handle.net/10603/14814.

Gansler, Jacques S., and William Lucyshyn. "Commercial-Off-the-Shelf (COTS):

Doing It Right." *Report.* University of Maryland, School of Public Policy. September 2008. 84. www.dtic.mil/dtic/tr/fulltext/u2/a494143.pdf

Goodwin, Tom. "The Battle is for the Customer Interface." *TechCrunch.* Accessed 17 September 2018. http://social.techcrunch.com/2015/03/03/in-the-age-of-disintermediation-the-battle-is-all-for-the-customer-interface/.

Gorman, Siobhan, Dreazen, Yochi J, and Cole, August. "Insurgents Hack U.S. Drones." *Wall Street Journal*, 18 December 2009, sec. US. http://www.wsj.com/articles/SB126102247889095011.

"Green Ship Technology Book: Existing Technology by the Marine Equipment Industry: A Contribution to the Reduction of the Environmental Impact of Shipping." European Marine Equipment Council., April 2010. https://www.oecd.org/sti/ind/48365856.pdf

Grevatt, Jon. "A Great Leap Forward." *Jane's Defence Weekly*, 03 May 2017. http://janes.ihs.com.inelibrary.remotexs.in/DefenceWeekly/Display/jdw65559-jdw-2017.

Habara, Keiji. "Maritime Policy in Japan" vol. 1 no. 1 (March 2011), 65–84.

Harish GK, and Singh, Prashant. "Shipbuilding: A Larger National Perspective." *New Delhi: Vivekananda International Foundation*, (January 2020). https://www.vifindia.org/sites/default/files/shipbuilding-a-larger-national-perspective.pdf.

He-rim, Jo. "Hyundai Heavy Industries Uses AI to Create 'Smart' Vessels." *The Korea Herald*, 13 January 2020, online edition, sec. Industry. http://www.koreaherald.com/view.php?ud=20200113000667.

Jeon, Maro, Jinmo Park, and Joohyun Woo. "Development of HHI's advanced Navigation Assistance System for Safe Voyage." *12th IFAC Conference on Control Applications in Marine Systems, Robotics, and Vehicles CAMS 2019*, vol. 52, no. 21 (01 January 2019), 111–13. https://doi.org/10.1016/j.ifacol.2019.12.292.

Jung, James. "Shipbuilding to Benefit from KT, Hyundai Heavy's 5G Tech." *KoreaTech Today*, 16 December 2019. https://www.koreatechtoday.com/shipbuilding-to-benefit-from-kt-hyundai-heavys-5g-tech/.

Kalouptsidi, Myrto. "China's Shipbuilding Industry: Measuring the Effect of Industrial Policy." *LSE Business Review*, 15 April 2019. https://blogs.lse.ac.uk/businessreview/2019/04/15/chinas-shipbuilding-industry-measuring-the-effect-of-industrial-policy/.

Lee, Dave. "Self-Navigating Cargo Ships 'by 2025.'" *BBC News*, 09 June 2017, sec. Technology. https://www.bbc.com/news/technology-40219682.

Lim, Chang-won. "Hyundai Shipyard Applies Autonomous Sailing Technology to Bulk Carrier." *Aju Business Daily*. 09 April 2020, sec. Industry. http://www.ajudaily.com/view/20200409173629074.

Lorge, Peter. "Water Forces and Naval Operations," A *Military History of China*, eds. David Graff and Robin Higham. Boulder: Westview Press, 2002.

Lukasiak, Lidia, and Andrzej Jakubowski. "History of Semiconductors." *Journal of Telecommunications and Information Technology* vol. 1, no. 1. (2010). 9. https://scholar.google.co.in/scholar_url?url=https://yadda.icm.edu.pl/baztech/element/bwmeta1.element.baztech-article-BATA-0008-0020/c/httpwww_itl_waw_plczasopismajtit201013.pdf&hl=en&sa=X&ei=H-O-YcvBLcKM6rQPhYCH8AI&scisig=AAGBfm0RCDHBJJfZhfRbTEAta-YgdqtjWQ&oi=scholarr

Mackinder, Halford J. "The Round World and the Winning of the Peace." *Foreign Affairs* vol. 21, no. 4 (1943), 595–605. https://doi.org/10.2307/20029780.

MarEx. "Shipbuilder Looks to Internet of Things for Future Business." *The Maritime Executive*, 07 June 2016. https://www.maritime-executive.com/article/shipbuilder-looks-to-internet-of-things-for-future-business.

Marr, Bernard. "Rolls-Royce And Google Partner to Create Smarter, Autonomous Ships Based On AI And Machine Learning." *Forbes*. Accessed 23 October 2018. https://www.forbes.com/sites/bernardmarr/2017/10/23/rolls-royce-and-google-partner-to-create-smarter-autonomous-ships-based-on-ai-and-machine-learning/.

Medeiros, Evan S, Roger Cliff, Keith Crane, and James C Mulvenon. "China's Shipbuilding Industry." *A New Direction for China's Defense Industry*. Rand Corporation, 2005. 109-154. https://www.jstor.org/stable/10.7249/mg334af.10.

Mes, Bjorn. "Virtual and Augmented Reality in Shipbuilding." *Damen Magazine*. Accessed 05 October 2018. https://magazine.damen.com/innovation/virtual-and-augmented-reality-in-shipbuilding/.

MI News Network. "DSME Solidifies Position as the World's Leading Smart Ship Building Leader." *Marine Insight*. 22 May 2020. https://www.marineinsight.com/shipping-news/dsme-solidifies-position-as-the-worlds-leading-smart-ship-building-leader/.

Moody, R Adam. "Reexamining Brain Drain from the Former Soviet Union." *The Non-proliferation Review* vol. 3, no. 3 (September 1996): 92–97. https://doi.org/10.1080/10736709608436643.

Moore, G E. "Progress in Digital Integrated Electronics." *IEEE Solid-State Circuits*

Society Newsletter vol. 11, no. 3 (September 2006), 36–37. https://doi.org/10.1109/N-SSC.2006.4804410.

Moreland, W. H. "The Ships of the Arabian Sea about A.D. 1500." *Journal of the Royal Asiatic Society of Great Britain and Ireland*, vol. 1, no. 1 (1939): 63–74.

———. "The Ships of the Arabian Sea about A.D. 1500 (Concluded from p. 74)." *Journal of the Royal Asiatic Society of Great Britain and Ireland*, vol. 2, no. 2 (1939): 173–92.

Murray, Geoffrey. "China Charts Course into LNG Shipbuilding - Global Times." *Global Times*. 02 October 2014. http://www.globaltimes.cn/content/841637.shtml.

Nan, Zhong. "Building a New Marine Economy - China - Chinadaily.Com.Cn." *China Daily*. 23 May 2017. http://www.chinadaily.com.cn/kindle/2017-07/23/content_30215621.htm.

Nath, Rajeshwar. "Shipbuilding in India." *Indian Defence Review* vol. 22, no. 2 (Apr-Jun, 2011). http://www.indiandefencereview.com/news/shipbuilding-in-india/.

Navab, N. "Developing Killer Apps for Industrial Augmented Reality." *IEEE Computer Graphics and Applications* vol. 24, no. 3 (May 2004): 16–20. https://doi.org/10.1109/MCG.2004.1297006.

"Navy to Accelerate LDUUV and XLUUV Acquisitions." *Ocean News and Technology*, 20 July 2018. https://www.oceannews.com/news/defense/navy-to-accelerate-lduuv-and-xluuv-acquisitions.

Nelson, Bryan. "10 Ways Graphene Could Change the World." *Mother Nature Network*. Accessed 16 October 2018. https://www.mnn.com/green-tech/research-innovations/stories/10-ways-graphene-could-change-the-world.

Nurkin, Tate. "China's Advanced Weapons Systems," *Jane's by HIS Markit*, 12 May 2018. https://www.uscc.gov/research/chinas-advanced-weapons-systems.

Organski, AFK. "Power Transition." *International Encyclopedia of the Social Sciences*. accessed 08 March 2018. http://www.encyclopedia.com/social-sciences/applied-and-social-sciences-magazines/power-transition.

Pak, Ki Hyuk. "Outcome of Land Reform in the Republic of Korea." *Journal of Farm Economics* vol. 38, no. 4 (1956): 1015–23. https://doi.org/10.2307/1234244.

Palais, James B. "A Search for Korean Uniqueness." *Harvard Journal of Asiatic Studies* vol. 55, no. 2 (1995): 409–25. https://doi.org/10.2307/2719348.

Palmisano, Samuel J. "The Globally Integrated Enterprise." *Foreign Affairs* vol. 85, no. 3 (2006): 127. https://doi.org/10.2307/20031973.

Panigrahi, Sumanta. "Asian Shipbuilding: A Dynamic Market." *GTR Export Finance Supplement.* (2014). http://www.citigroup.com/transactionservices/home/trade_svcs/docs/asian_shipbuilding.pdf.

Park, Kyunghee. "Blockchain Is about to Revolutionise the Shipping Industry." *The Economic Times*, 20 April 2018. https://economictimes.indiatimes.com/industry/transportation/shipping-/-transport/blockchain-is-about-to-revolutionize-the-shipping-industry/articleshow/63830173.cms.

———. "Labour-Intensive Shipbuilding Industry Employing Robots to Cut Costs." *Business Standard India*, 16 April 2018. https://www.business-standard.com/article/international/labour-intensive-shipbuilding-industry-employing-robots-to-cut-costs-118041600155_1.html.

Parkinson, Bradford W, and Stephen T Powers. "Part 1: The Origins of GPS, and the Pioneers Who Launched the System." *GPS World*, 02 May 2010. https://www.gpsworld.com/origins-gps-part-1/.

———. "Part 2: The Origins of GPS, Fighting to Survive." *GPS World*, 01 June 2010. https://www.gpsworld.com/origins-gps-part-2-fighting-survive/.

Quinones, C. Kenneth. "The Impact of the Kabo Reforms upon Political Role Allocation in Late Yi Korea, 1884-1902." *Occasional Papers on Korea*, no. 4 (September 1975): 1–18.

Randhawa, BS. "Indian Shipbuilding: Key to Maritime and Economic Security." *Indian Defence Review* vol. 25 no.1, (Jan-Mar, 2010). http://www.indiandefencereview.com/spotlights/indian-shipbuilding-key-to-maritime-and-economic-security/.

Rao, S.R. "Shipping and Maritime Trade of Indus People." *Expedition,* vol. 7 no.3, 1965.

Rawung, R H, and Putrada, A G. "Cyber Physical System: Paper Survey." *International Conference on ICT For Smart Society (ICISS)*, 273–78, 2014. https://doi.org/10.1109/ICTSS.2014.7013187.

Sakhuja, Vijay. "Autonomous Ship Operations in Need of New Regulatory Regime." *Society for the Study of Peace and Conflict*, 30 August 2018. https://www.sspconline.org/index.php/opinion-analysis/autonomous-ship-operations-need-new-regulatory-regime-thu-08302018.

Sarangi, Subhasish. "Unpacking SAGAR (Security and Growth for All in the Region)." *United Service Institution of India, Occasional Paper*, vol. 2 (2019), 12.

Satell, Greg. "The Industrial Era Ended, and So Will the Digital Era." *Harvard Business Review*, 11 July 2018. https://hbr.org/2018/07/the-industrial-era-ended-and-so-will-the-digital-era.

Savvides, Nick. "Outlook 2018: Asian Shipyards to Embrace Innovation in 2018." *IHS Markit Safety at Sea*, 28 December 2017. https://safetyatsea.net/news/2017/outlook-2018-asian-shipyards-to-embrace-innovation-in-2018/.

Schuman, Michael. "More and More Families Are Joining the Global Middle Class." *US News & World Report*. Accessed 10 April 2018. https://www.usnews.com/news/best-countries/articles/2018-01-23/asian-consumers-becoming-most-powerful-economic-force-in-world.

Segan, Sascha. "What Is 5G?" *PCMag India*, 16 April 2019. https://in.pcmag.com/cell-phone-service-providers/104415/what-is-5g.

Shin, Kyoung-Ho, and Paul Ciccantell. "The Steel and Shipbuilding Industries of South Korea: Rising East Asia and Globalization." *Journal of World-Systems Research* vol. 2 (January 1, 2009): 167–92. https://doi.org/10.5195/jwsr.2009.316.

Shkolnikov, Vladimir. "Potential Energy: Emergent Emigration of Highly Qualified Manpower from the Former Soviet Union." *Pardee RAND Graduate School*, 1994. https://www.rand.org/pubs/rgs_dissertations/RGSD113.html.

Siegel, J E, Kumar, S, and Sarma, SE. "The Future Internet of Things: Secure, Efficient, and Model-Based." *IEEE Internet of Things Journal* vol. 5, no. 4 (August 2018): 2386–98. https://doi.org/10.1109/JIOT.2017.2755620.

Singh, Ashish. "Cochin Shipyard Signs Contract to Build Electric Vessels for Norwegian Company." *The Daily Guardian*, 26 August 2020. https://thedailyguardian.com/cochin-shipyard-signs-contract-to-build-electric-vessels-for-norwegian-company/.

Sisinni, E, Saifullah, A, Han, S, Jennehag, U, and Gidlund, M. "Industrial Internet of Things: Challenges, Opportunities, and Directions." *IEEE Transactions on Industrial Informatics*, vol. 1, no. 1, (2018). https://doi.org/10.1109/TII.2018.2852491.

Smith, Anne Marie. "Phoenician Ships: Types, Trends, Trade and Treacherous Trade Routes." *Dissertation, University of South Africa*, 2012. http://uir.unisa.ac.za/handle/10500/10344.

Smyth, Russell, Xin Deng, and Junli Wang. "Restructuring State-Owned Big Business in Former Planned Economies: The Case of China's Shipbuilding Industry." *New Zealand Journal of Asian Studies*, vol. 6, no. 1 (June 2004): 30.

"South Korea Coordinates Autonomous Ship Efforts." *The Maritime Executive*, 17 June 2020. https://www.maritime-executive.com/article/south-korea-coordinates-autonomous-ship-efforts.

Stanic, Venesa, Marko Hadjina, Niksa Fafandjel, and Tin Matulja. "Toward Shipbuilding 4.0-an Industry 4.0 Changing the Face of the Shipbuilding Industry." *Brodogradnja,* vol. 69 (30 September 2018), 111–28. https://doi.org/10.21278/brod69307.

Tan, Liwei. "Made in China 2025 Policy: Maritime Equipment and High-Tech Ships" *Presented at the Maritime Business Day*, Espoo, Finland: 10 May 2017. https://www.slideshare.net/FinproRy/made-in-china-2025-policy-maritime-equipment-and-hightech-ships.

"The Challenges of Using New Materials in Shipbuilding." *Marine Offshore Technology*. Accessed 14 October 2018. http://www.marineoffshoretechnology.net/features-news/challenges-using-new-materials-shipbuilding.

"The Final Countdown." *The Economic Times*. 25 July 2019.

Tian C. "The Six Decades of Chinese Industry." *The Economic Observer*. 07 April 2009. http://www.eeo.com.cn/ens/news/.

Tovey, Alan. "Virtual Reality Warships: Why BAE Is Diving into 3D," *The Telegraph*, 05 November 2014. https://www.telegraph.co.uk/finance/newsbysector/industry/defence/11210224/Virtual-reality-warships-Why-BAE-is-diving-into-3D.html.

Will, Martin. "The Global Cryptocurrency Market Hit a New Record High above $700 Billion." *Business Insider*. Accessed 11 October 2018. https://www.businessinsider.com/bitcoin-price-global-cryptocurrency-market-capitalisation-january-3-2018-1.

"Virtual & Augmented Reality - Understanding the Race for the Next Computing Platform." *Global Investment Research. Goldman Sachs*, 13 January 2016. https://www.goldmansachs.com/insights/pages/technology-driving-innovation-folder/virtual-and-augmented-reality/report.pdf.

Walker, Jon. "Autonomous Ships Timeline - Comparing Rolls-Royce, Kongsberg, Yara and More." *Tech Emergence*, 29 May 2018. https://www.techemergence.com/autonomous-ships-timeline/.

Wallace, Benjamin. "The Rise and Fall of Bitcoin." *Wired*. Accessed January 2, 2021. https://www.wired.com/2011/11/mf-bitcoin/.

Walsh, Sue. "IBM, Maersk Roll Out Blockchain-Based Shipping Platform." *RT Insights*, 06 September 2018. https://www.rtinsights.com/ibm-maersk-roll-out-blockchain-based-shipping-platform/.

Werner, Ben, and Eckstein, Megan. "Palm-Sized 3D-Printed Part Represents Leap Forward in Shipbuilding." *USNI News*, 12 October 2018. https://news.usni.org/2018/10/12/palm-sized-part-represents-leap-forward-in-shipbuilding.

World Maritime News. "Samsung Heavy Joins Forces with MAN on Smart Ship Tech." *Offshore Energy*, 19 August 2019. https://www.offshore-energy.biz/samsung-heavy-joins-forces-with-man-on-smart-ship-tech/.

Yoon, Sukjoon. "Make Way for South Korea's Underwater Drones." *The Diplomat*, 19 February 2020. https://thediplomat.com/2020/02/make-way-for-south-koreas-underwater-drones/.

Zanero, Stefano. "Cyber-Physical Systems." *IEEE Computer Society*, vol. 50, no. 4 (April 2017), 14–16. https://doi.org/10.1109/MC.2017.105.

Zhu, Xiaodong. "Understanding China's Growth: Past, Present, and Future." *Journal of Economic Perspectives,* vol. 26, no. 4 (November 2012), 103–24. https://doi.org/10.1257/jep.26.4.103.

Online Sources/Websites

Abdullah, Zhaki. "Driverless Shuttle Trials Start on Sentosa; on-Demand Service for Public from 2019." *The Straits Times*, 05 June 2018. https://www.straitstimes.com/singapore/transport/driverless-shuttle-trials-to-start-on-sentosa-june-5-on-demand-for-public-from.

"ACTUV 'Sea Hunter' Prototype Transitions to Office of Naval Research for Further Development." Accessed 06 September 2018. https://www.darpa.mil/news-events/2018-01-30a.

Ancient China Facts. "Famous Ancient Chinese Ships, The Castle Ship Shipbuilding Techniques." Accessed 20 July 2018. http://www.ancientchinalife.com/famous-ancient-chinese-ships.html.

Bartlett, Paul. "Mixed Reality Technology to 'Disrupt' Shipbuilding and Maintenance." 20 June 2018. http://www.seatrade-maritime.com/news/europe/mixed-reality-technology-to-disrupt-shipbuilding-and-maintenance.html.

"Byzantine Empire - Ancient History". http://www.history.com/topics/ancient-history/byzantine-empire.

Chappine, Patricia. "The Second Industrial Revolution: Timeline & Inventions." *Study.com.* http://study.com/academy/lesson/the-second-industrial-revolution-timeline-inventions.html.

"Creation of the Bretton Woods System | Federal Reserve History." https://www.federalreservehistory.org/essays/bretton_woods_created.

"Designing and Building a Wooden Ship." http://penobscotmarinemuseum.org/pbho-1/ships-shipbuilding/designing-and-building-wooden-ship.

"DNV GL Technology Outlook 2025 – Shipping and Digitalization." Accessed 27 October 2018. https://to2025.dnvgl.com/shipping/digitalization/.

Duffie Jr., Warren. "Navy Developing Ship Coatings to Reduce Fuel, Energy Costs - Office of Naval Research," 21 June 2018. https://www.onr.navy.mil/en/Media-Center/Press-Releases/2018/Navy-Developing-Ship-Coatings.

Encyclopedia Britannica. "South Korea - Economic and Social Developments." Accessed 16 May 2020. https://www.britannica.com/place/South-Korea.

"Entrade, the Clean Energy Company." //www.schneider-electric.co.in/en/work/campaign/life-is-on/case-study/entrade.jsp.

Experts Exchange. "Processing Power Compared: Visualizing a 1 Trillion-Fold Increase in Computing Performance." Accessed 01 November 2018. https://pages.experts-exchange.com/processing-power-compared.

Freist, Roland. "General Electric Converts Diesel Ships into Smart Navy - Digital Twin," 26 February 2018. http://www.hannovermesse.de/en/news/general-electric-converts-diesel-ships-into-smart-navy-71809.xhtml.

"Global Claims Review - Liability in Focus." Loss trends and emerging risks for businesses. Munich, Germany: Allianz Global Corporate & Specialty, 01 March 2017. https://www.agcs.allianz.com/assets/PDFs/Reports/AGCS-Global-Claims-Review-2017.pdf.

"GPS.Gov: GPS Accuracy." Accessed 17 May 2019. https://www.gps.gov/systems/gps/performance/accuracy/.

Grey, Eva. "3D Printing: Rising to the Challenge in Ship Design." *Ship Technology*, 26 October 2015. http://www.ship-technology.com/features/feature3d-printing-rising-to-the-challenge-in-ship-design-4672912/.

Grieves, Michael. "Digital Twin: Manufacturing Excellence through Virtual Factory Replication," (2014). http://innovate.fit.edu/plm/documents/doc_mgr/912/1411.0_Digital_Twin_White_Paper_Dr_Grieves.pdf.

Hays, Jeffrey. "Economic History after World War II in Japan, South Korea and South East Asia." Accessed 12 August 2018. http://factsanddetails.com/asian/cat62/sub408/item2560.html.

———. "Minoans (3000 B.C. to 1400 B.C.): Their Art, Culture and Religion and the Thera Eruption." Accessed 13 July 2018. http://factsanddetails.com/world/cat56/sub366/item2043.html.

Hribernik, Karl. "Industry 4.0 in the Maritime Sector." Bremer Institute for Production and Logistics, Bremen, Germany, 13 April 2016, 22. http://www.mlit.go.jp/common/001127983.pdf

Inc.com. "Industry Life Cycle," Accessed 31 July 2016. http://www.inc.com/encyclopedia/industry-life-cycle.html.

Index AR Solutions. "Digital Reality Technology: Empowering Today's Knowledge

Workers," 02 August 2018. https://www.indexarsolutions.com/digital-reality-technology-empowers-knowledge-workers/.

"Internet of Everything - Capabilities for the US Navy." White Paper. Cisco, 2015. https://www.cisco.com/c/dam/en_us/solutions/industries/us_govern ment/resources/navy-ioe-wp1c.pdf.

"Japan: IHI Marine United, Universal Shipbuilding Merger Creates Giant," 04 January 2013. https://worldmaritimenews.com/archives/114922/japan-ihi-marine-united-universal-shipbuilding-merger-creates-giant/.

Kelly, Julian. "A Preview of Bristlecone, Google's New Quantum Processor." *Google AI Blog,* 05 March 2018. http://ai.googleblog.com/2018/03/a-preview-of-bristlecone-googles-new.html.

"Milestones: 1945–1952 - Office of the Historian." Accessed 23 March 2018. https://history.state.gov/milestones/1945-1952/kennan.

Nakamoto, Satoshi. "Bitcoin: A Peer-to-Peer Electronic Cash System," Accessed 20 September 2018. https://bitcoin.org/bitcoin.pdf

Naval Technology. "Sejong the Great Class / KDX-III Class Destroyer." Accessed 11 July 2020. https://www.naval-technology.com/projects/sejongthegreat classd/.

Nishioka, Yasuyuki. "What's IVI? – Industrial Valuechain Initiative." Accessed 22 July 2019. https://iv-i.org/wp/en/about-us/whatsivi/.

Nye, Joseph S Jr. "Understanding 21st Century Power Shifts." Accessed 11 April 2018. http://www.europeanfinancialreview.com/?p=2743.

Polzer, Jorg. "Blockchain Technology: A Game Changer in Shipbuilding Industry," 26 January 2018. https://www.linkedin.com/pulse/blockchain-technology-game-changer-shipbuilding-industry-j%C3%B6rg-polzer.

Riviera Maritime Media. "Japan Prominent in Latest LNG Fleet Surge," 11 May 2018. https://www.rivieramm.com/opinion/opinion/japan-prominent-in-latest-lng-fleet-surge-24777.

Riviera Maritime Media. "Kawasaki Ship Designs Support Japan's Hydrogen-Society Plans," 07 February 2017. https://www.rivieramm.com/news-content-hub/news-content-hub/kawasaki-ship-designs-support-japans-hydrogen-society-plans-29850.

Rouse, Margaret. "What Is AI (Artificial Intelligence)? - Definition from WhatIs.Com." *Search Enterprise AI.* Accessed 29 September 2018. https://searchenterpriseai.techtarget.com/definition/AI-Artificial-Intelligence.

"Shuttleworth Design - Mayflower Autonomous Research Ship," 08 August 2016. http://www.shuttleworthdesign.com/gallery.php?boat=MARS.

"Six Theories about How 3D Printing Will Change Logistics." *AEB GmbH, Stuttgart,* (November 2017). www.aeb.com.

"South Korea Invites Local Firms to Design Aegis-Equipped Destroyer." Accessed 11 July 2020. https://www.defenseworld.net/news/27102/South_Korea_ Invites_Local_Firms_to_Design_Aegis_equipped_Destroyer#.XwnWb SgzY2x.

"South Korea Submarine Capabilities." Accessed 11 July 2020. https:// www.nti.org/analysis/articles/south-korea-submarine-capabilities/.

"The CANES Evolution." Accessed 03 October 2018. https://www.public. navy.mil/spawar/PEOC4I/Pages/CANESTest.aspx.

"Truman Doctrine Is Announced - Mar 12, 1947." Accessed 11 August 2018. http://www.history.com/this-day-in-history/truman-doctrine-is-announced.

Tvete, Hans Anton. "The ReVolt - A New Inspirational Ship Concept." *DNV GL,* 08 August 2016. https://www.dnvgl.com/technology-innovation/revolt/ index.html.

UN News. "UN Projects World Population to Reach 8.5 Billion by 2030, Driven by Growth in Developing Countries," 29 July 2015. https://news.un.org/ en/story/2015/07/505352-un-projects-world-population-reach-85-billion-2030-driven-growth-developing.

Walendowski, Jacek, Kroll, Henning and Schnabl, Esther. "Industry 4.0, Advanced Materials (Nanotechnology)," Accessed 15 October 2018. https:// ec.europa.eu/growth/tools-databases/regional-innovation-monitor/sites/ default/files/report/RIM%20Plus_Industry%204.0%2C%20Advanced%20 Materials%2 0%28Nanotechnology%29_Thematic%20paper.pdf

"What is Quantum Computing? - IBM Q - US." Accessed 29 October 2018. // www.research.ibm.com/ibm-q/learn/what-is-quantum-computing.

World Maritime News. "China's 1st Smart Ship Makes a Debut," 07 December 2017. https://worldmaritimenews.com/archives/237342/chinas-1st-smart-ship-makes-a-debut/.

Zhihao, Zhang. "Finding Paves Way for Even Better Computers." Accessed 18 August 2018. https://www.chinadaily.com.cn/a/201808/18/WS5b77659da3 10add14f386736.html.

Index